P9-ELW-104

This Old House Heating, Ventilation, and Air Conditioning

Illustrations: Michael Blier
Design: Chris Pullman/WGBH
Illustration editor: Pamela Hartford
Additional drawing: John Murphy

This Old House Heating, Ventilation, and Air Conditioning

A Guide to the Invisible Comforts of Your Home

Richard Trethewey with Don Best

Little, Brown and Company
Boston New York Toronto London

Also in this series:
This Old House Kitchens: A Guide to Design and Renovation
This Old House Bathrooms: A Guide to Design and Renovation

First Edition

Library of Congress Cataloging-in-Publication Data

Trethewey, Richard.
This Old House heating, ventilation, and air conditioning : a guide to the invisible comforts of your home / Richard Trethewey with Don Best. — 1st ed.
p. cm.
Includes index.
ISBN 0-316-85271-6 (hc)
ISBN 0-316-85272-4 (pbk.)
1. Heating. 2. Ventilation. 3. Air conditioning. I. Best, Don, 1949– . II. Title.
TH7011.T74 1994
697 — dc20 93-20584

10 9 8 7 6 5 4 3 2 1

RRD-OH

Published simultaneously in Canada by Little, Brown & Company (Canada) Limited

Printed in the United States of America

Author's Note

The point of this book is to provide you with all the information you need to make smart decisions about heating, ventilation, and air conditioning in your home. It is a subject many homeowners elect not to pay attention to until something breaks down. Much of the equipment is buried in the walls, floors, and basements of our homes, and what has been visible — bulky radiators or rattling window air conditioners — has often not been a joy to eye or ear. But our basic comfort in our homes is at stake in these systems. Industrial research keeps offering us new alternatives, and with them the challenge of measuring the expense of maintaining the old against the cost of installing and operating the new. The equipment, fortunately, is getting smaller, and easier to look at without scowling.

I might still be helping to manage a four-generation-old family plumbing business in Massachusetts except that in 1978 Russell Morash at WGBH in Boston, having established a now venerable gardening series for public television, "The Victory Garden," created a series about home renovation. Russ called the Boston Gas Company to get some recommendations about plumbing contractors. Our family business, Trethewey Brothers, was on the list. Another plumber was enlisted first, but he pulled out shortly before the first show was to be videotaped, and my father, Ron Trethewey, was called.

It was certainly not a promising invitation financially. The show, to be called "This Old House," had brought a tired old Victorian wreck in Dorchester, Massachusetts, for $17,000 and planned to renovate it on a $30,000 budget; the renovation would include a new heating and water system, all new electrical wiring, a new roof, basic structural repair, and a complete cosmetic overhaul! But my father, who was a most knowledgeable plumbing contractor, was eager to do it because he felt the television exposure might improve the public's perception of plumbers and their work. What he didn't reckon was that television cameras could be so daunting. I worked on the plumbing crew with him at the Dorchester house, which blossomed into a Victorian beauty, and at the end of the first season acceded to my father's plea to take his place on camera.

The rest is a bit of television history. "This Old House" has become the most popular half-hour program on public television. In my area, we have introduced viewers to passive solar heating, radiant underfloor heating, split system air conditioning, and many other improved technologies. Viewers know how frequently, at the beginning of renovation projects, I have found existing heating, water delivery, and air conditioning systems to be among the most antiquated and inefficient aspects of the house. Getting the best affordable comfort has become the focus of my work, and I pursue the subject both on the television series and in other forums with homeowners and plumbing contractors.

Working with executive producer and friend Russ Morash has been one of the greatest pleasures of my life. Add to that a long association with Norm Abram, who has mastered carpentry and virtually every other aspect of home construction and renovation; and with the lively host of "This Old House," Steve Thomas. In my seminars and personal appearances I always pay tribute to Tom Silva and his crew, who have worked on many "This Old House" renovations, and to the splendid production crew, especially cameraman Dick Holden and producer Bruce Irving.

Preparing a book requires as elaborate a team as producing a television series. Don Best has been extremely diligent in helping me research and write the text, and I am grateful for his competence and his friendship. Michael Blier, assisted by John Murphy, has transformed the artifacts of my trade into art. Pamela Hartford has directed the illustrations and kept everyone on course. Chris Pullman has demonstrated his wizardry of book design once again. And William Phillips and Catherine Crawford at Little, Brown have provided every publishing support.

A final word of thanks to my wife, Christine, and my sons, Ross and Evan, who accepted many hours of my working on this book in my barn-office that I otherwise would have spent lovingly with them.

Richard Trethewey

This Old House Heating, Ventilation, and Air Conditioning

The Hidden Keys to Comfort

I've come to realize over the years that the work I do, so far as most folks are concerned, isn't very sexy. It's pretty common for people to wax romantic about their French doors and fluted columns, or to fall in love with a graceful old staircase. But I'm still waiting for the day when a homeowner rushes up to me and exclaims, "Wow, Richard, that new boiler you put in is really sexy!"

Looks good, they might say. Or nicely installed. But *never* sexy.

It is one of the great ironies — and flaws — in the construction and real estate business that so much of the home's value is reckoned on the surface, while its hidden qualities — the very things that determine people's comfort — go almost unexamined.

An alarming number of home builders, remodelers, and real estate agents still live by the old credo "You can't sell hidden value." What they mean by that, of course, is that their customers still judge a house mostly by the shape and condition of its skin, and don't pay much attention to the bones and muscle.

The Wickwires' old post-and-beam barn in Concord, Massachusetts, was almost completely reframed and then encased with superinsulating stress-skin panels. To retain the structure's historical flavor, we finished the exterior with wood clapboards.

Though Lynn and Barbara wanted a fireplace for symbolic warmth (below), the heart of the comfort system is a high-efficiency boiler with weather-responsive controls (facing page). Note the series of tubes rising from the boiler's twin manifolds. The upper manifold supplies warm water to the home's baseboard radiators and towel warmers while the lower manifold supplies the radiant floor heating system.

While cameraman Dick Holden (left) frames the shot, producer Russ Morash (right) gives me some last-minute direction on our project house in Wayland, Massachusetts.

But I'm here to tell you a secret: *comfort is invisible.* It's hidden *behind* the fancy wainscot, in the construction of the wall and the value of the insulation. It's hidden in the pipes and ductwork and wiring. It's hidden down in the basement and up in the attic — the very last places the real estate agent wants to show.

I can't resist telling a story here about Spencer Tracy, who must have known a lot about hidden values. As the story goes, Tracy had granted an exclusive interview to a rather wet-behind-the-ears reporter, who could barely get his questions out because he was so nervous in the presence of the great actor. Finally, to set the kid at ease, the inimitable Tracy leaned over toward him and said, "*Acting* is not an important job in the scheme of things. *Plumbing* is."

If I do nothing but help you establish smart priorities for your current or future home, then I will have accomplished a great deal.

Chandeliers or Thermostats?

There's an instructive little game that host Steve Thomas and I sometimes play on "This Old House." He'll ask me, on camera, what I would advise doing with an old plumbing or heating system "if money were no object." This gives us the chance to explore the outer limits of remodeling, where Rolls-Royce heating systems and Cadillac plumbing dwell, where labor costs are no concern, where no one blinks an eye at ripping out the old and buying new.

The truth, of course, is that money is *always* an object. And budgets are important even to the wealthy.

Steve and I always finish our game with a much tougher and down-to-earth question: "And what would you do if money *were* an object?"

"Well," I always say, "It *depends* . . ."

This is the point, both in the show and in your remodeling projects at home, where reality sets in. Suddenly you're faced with an almost endless confusion of options and tradeoffs. And with each comes a marketing organization to convince you of its merit.

From doorbells to dormers to ductwork, everyone is out there vying for your dollar. When you consider the fact that Americans spend more than $110 billion a year on home renovation projects, you begin to understand why your mailbox is crammed full of catalogs, magazines, and mailers, all bent on telling you how to spend your money.

In my view, it's the home's heating, ventilation, and air conditioning (HVAC) systems, along with its plumbing, that set the foundation on which the family's physical comfort and health depend. These same systems also affect the family's wealth, not just in the price of their purchase, but in the ongoing cost of operating and maintaining them.

To those who suggest that veneer and fancy furnishings should come first in your remodeling agenda, and that the dull, mechanical, invisible projects can wait until later, I pose the following questions:

What good does it do you to have elaborate stained glass in the windows if they're steamed up with condensation half the time? Or to have Italian tile on your bathroom floor if it frostbites your toes come winter? Or to sit in your hot tub contemplating an enormous electric bill? These are the kind of ironic situations you have to face when comfort is only skin deep.

Before you label me an incurable pragmatist, let me tell you about my other side. I like gingerbread trim and brass doorknobs, Palladian windows and marble mantels, deep pile carpets and decorative faucets. I think being frivolous can be fun — just as long as it doesn't undermine comfort, health, or wealth.

One of my favorite all-time projects on "This Old House" was the post-and-beam barn-style house that we erected in Concord, Massachusetts. Some of you will remember the memorable stress-skin panels we used to superinsulate the walls, the in-floor hydronic heating, and various other measures we took to make that house truly comfortable and energy-efficient. I think it came as a shock to some of the purists on the crew and in the TV audience when the owners, Lynn and Barbara Wickwire, decided they wanted the "coziness and romance" of a fireplace.

Take it from me: their decision wasn't made lightly. They understood from the start that fireplaces tend to be losers as far as heating is concerned. So when they chose "romance," they tempered the decision with practicality by making sure their new fireplace had an effective damper, airtight doors, and an outside source of combustion air. They ended up with a fireplace that's probably a break-even proposition as far as energy is concerned — maybe even a small net plus.

This is what I'd call a "smart" frivolity, because Lynn and Barbara understood the tradeoffs *before* they made their choice. "Stupid" frivolities are the ones you buy with your eyes closed, only to have them rudely opened later on.

It's important to note that one person's "luxury" is very often another person's "necessity." Installing central air conditioning in Buffalo, New York, seems pretty frivolous to me, but in El Paso, Texas, I'd put it near the top of my list.

As we'll discover in chapters to come, setting priorities in heating, ventilation, and air conditioning work can sometimes be as much an art as a science. Just as people are different from one another, so are their houses, lifestyles, and budgets — and perhaps most important, the climates in which they live.

Changing the Way We Build

While I was in Germany attending a trade show on heating and cooling technology, I discovered that Europeans tend to have a remarkably different attitude toward their houses than we have in this country. Though I'm going to get hate mail from every quality builder in the United States for saying this, I think modern European houses tend to be better built and more energy-efficient than most typical U.S. construction. In fact, the reason I was in Germany for that show (rather than Cleveland or St. Louis) was because the Europeans, and particularly the Scandinavians, are setting the pace with some of their energy-efficient construction techniques and HVAC technologies.

There are powerful cultural, historical, and geographical reasons for this. For one thing, Europeans don't have the wanderlust that has been so much a part of our own culture. It's not uncommon for a European family to inhabit the same house for four or five generations, which means they have the motive and the time to invest in quality. Because we Americans, by comparison, move so frequently, we tend to think of houses as transitory habitats. With that in mind, some builders throw up ticky-tack houses with tin-pan heating and cooling systems. Perfect dwellings, I suppose, if you happen to be a gypsy.

Another reason the Europeans are so keen on quality homes and high-efficiency HVAC systems is because they were faced with limits on their natural resources a long, long time ago. The prices they pay for building materials, heating fuels, electricity, and water fairly accurately reflect the limitations of their natural resource base and the cost of imports.

Only now are Americans waking up to the fact that our oil supplies *are* being exhausted, that our forests aren't infinite, that clean air and water can't be taken for granted.

People are beginning to understand that the way they use fuel and water at home affects not only their own comfort, health, and economy, but everyone else's. Such seemingly distant and unrelated problems as the depletion of the earth's ozone layer, the push to drill for oil in our national parks, the growing shortage of pure drinking water in some states and communities, air pollution from power plants, cars, and homes, and our nation's ongoing deficit in international trade have finally come home to roost in our living rooms.

Even as we turn "greener" in the way we think and act, our society is also getting "grayer." And the older we get, the less inclined we are to move.

Back in the 1950s and '60s, an incredible 20 percent of our population moved every year. But the Census Bureau says that by the time the 1980s were over, our annual mobility rate had fallen to 18 percent, and that it's likely to slow even further during the '90s.

Compared to the Europeans, of course, we're still a nation of overnight campers. But the fact that we're staying in our houses longer — about ten years on average now — means that we're increasingly willing to invest in long-term comfort, health, and economy. And it means that we're going to be demanding more quality — especially *hidden* quality — in the houses we build. As people begin to reflect these changing values in the marketplace, builders, remodelers, and real estate agents are sure to get the message.

Rethinking Our Standards

Everywhere I look I see signs of a changed and changing America that is refocusing its attention on quality instead of quantity and underlying values instead of glitz and hype. This is true even of our grand old institutions, which are usually slow to change.

For example, the National Association of Realtors (NAR) is pushing for new regulations that would force home sellers to disclose the *true* condition of their house — especially any hidden defects — before it goes up for sale. Trust me, the NAR isn't doing this out of the goodness of their heart or because they covet more regulation. They want new disclosure laws enacted in all fifty states to protect themselves and their member Realtors from lawsuits filed by angry buyers who feel they've been cheated.

Examples abound of people moving into their dream home only to discover that the furnace is in fact a lemon, or that the plumbing, which dates back to the Taft administration, is leaky and full of lead, or that underneath that smart veneer, the studs are starting to rot because some fool put the vapor barrier in on the wrong side of the insulation.

Right now, sixteen states• have mandatory disclosure laws. In the other thirty-four states, where disclosure is either voluntary or nonexistent, the law of the jungle still prevails: "Let the Buyer Beware." Still, I'm encouraged that sellers, Realtors, and buyers are going to be taking a closer look at those hidden qualities that impart real value to a house.

I'm also encouraged to see that the federal government and the building code organizations in the United States are moving toward higher standards. The problem is, it takes time for these changes to trickle down to the state and local level, and even then enforcement can vary from fairly strict to none at all.

While most states now have some kind of energy code for new houses — prescribing minimum insulation levels and other standards — they're typically ten to fifteen years out of date. A study by the Alliance to Save Energy, a nonprofit coalition of business, government, environmental, and consumer leaders, found that as of September 1991, two-thirds of the states, producing more than half the nation's housing stock, had building energy codes that don't even come close to modern technical and economic standards.

Alliance senior program manager Bion Howard, a specialist in building science, says that if the states were uniformly to accept the Model Energy Code (a voluntary national standard), it would be a tremendous boon to homeowners, not to mention the dividends it would pay toward an improved national economy and a cleaner environment.

"Our research shows that building a new house in accordance with the Model Energy Code — in those thirty-four states that aren't complying — would add $1,100 to the price of a single-family house," says Bion. "But when you roll that extra money into a thirty-

• Alaska, California, Illinois, Indiana, Kentucky, Maine, Maryland, Michigan, Mississippi, New Hampshire, Ohio, Rhode Island, South Dakota, Texas, Virginia, and Wisconsin.

year mortgage, it comes out to about $8 a month, while the energy savings achieved are $15 a month. In other words, the homeowner has a positive cash flow on the deal from day one, not to mention a house that's a lot more comfortable to live in."

If the Model Energy Code were adopted nationwide, it would save the United States 5.4 billion cubic feet of natural gas, 600 million kilowatts of electricity, and nearly a million gallons of fuel oil in the first year alone, Bion reckons. It would also eliminate 653,000 tons of energy-related air pollution, mostly in the form of carbon dioxide.

The push to raise the quality of America's housing stock and improve its energy efficiency is also involving agencies of the federal government. The U.S. Department of Energy (DOE) is helping industry to develop guidelines for a Home Energy Rating System that will give home builders, owners, real estate agents, mortgage bankers, and buyers reliable information on the energy performance of new and existing houses — the same way that cars are assigned a miles-per-gallon rating.

Some states, local communities, and utilities are already experimenting with auditing programs that rate houses with points, stars, or seals of approval according to their energy performance. But the DOE program will go further, creating a voluntary national standard.

"When you have a house audited and rated, it gives you a clearer picture of where you stand, and points out the most cost-effective ways to make improvements," says Mary-Margaret Jenior, head of DOE's Office of Buildings Research. "From a buyer's point of view, a good, objective rating can quickly separate the hogs from the Thoroughbreds."

The federal government has also enacted new legislation that should make it easier for consumers to get an Energy Efficient Mortgage, or EEM. These special mortgages enable home buyers and owners to roll the cost of energy improvements into the cost of their first mortgage or refinancing. By singling out those improvements that are most cost-effective, homeowners can end up saving more money in energy each month than the incremental rise in their mortgage payment.• EEMs can also be used to help buyers of energy-efficient homes qualify for a larger loan.

• Energy Efficient Mortgages are available through the Federal Housing Administration (FHA), Department of Veterans Affairs (VA), Federal National Mortgage Association (Fannie Mae), and the Federal Home Loan Mortgage Corporation (Freddie Mac). These secondary mortgage institutions set the guidelines for purchasing EEMs from banks, savings and loans, and other primary lenders that work directly with consumers. Until the financial community becomes more familiar with EEMs, you may have to shop around to find a lender who's read the fine print and stands ready to help.

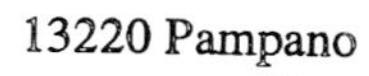

13220 Pampano   Little Rock, AR 72204

Square Footage: 2115

FEATURES

Ceiling	R-35
Floor	R-11
Walls	R-24
Windows	Low E (double pane)
Air leakage	0.35 Natural air changes/hr
Solar gain area	10% of floor space
Water Heater	Fuel: gas Location: Garage Efficiency: Level D: Tank and pipes wrapped, heat trap
Space Heating	Lennox 70% Efficient Secondary: Wood Stove
Air Conditioning	Lennox 12 SEER
Ductwork	Location: conditioned space Insulation: R-5
Other devices	Continuous soffit/ridge vents Energy Landscape Whole house fan

RATING:

★★★★★

91.8

July 25, 1988
Date of Rating

Ronald E. Hughes
Energy Rater

Energy Rated Homes of Arkansas

Energy Label™

Labels like this, which provide details about the energy efficiency of a home, will become more and more common in the years ahead. (See text on page 9.) This particular house, in Little Rock, Arkansas, was assigned a five-star rating, which means it's a real champion when it comes to energy.

According to the plans and promotional campaign laid out under the National Affordable Housing Act of 1990, and further strengthened by the National Energy Policy Act of 1992, the Home Energy Rating System and EEMs are supposed to be available to everyone in the United States by 1995. This seems to be one of those rare deals where everybody wins. Homeowners get a better, more comfortable house with lower operating costs; bankers and builders will have a larger market to serve, since EEMs make housing more affordable; even the utilities should be pleased, since most of them are trying to shave their peak loads through energy conservation.

As a cautionary tip, I should point out that an energy-efficient house isn't always a comfortable or healthy house. It's possible, for example, for a house with a five-star energy rating to be noisy, to run hot and cold, to have poor air quality or lousy plumbing. Still, the energy efficiency that's built into a home, along with the quality of its HVAC systems, are generally good indicators of comfort, health, and economy. If a builder is conscientious in insulating and sealing the home's frame, and installing the right HVAC systems, it's a good bet that he's followed through on the electrical work, plumbing, and other details.

I'll have more to say about energy audits, home rating systems, and EEMs later on. My point here is that when government agencies, mortgage bankers, and other varieties of dinosaurs start to move, you can be sure the world is changing. By the time the '90s are over, I may have a whole list of clients who think heating, ventilation, and air conditioning are the sexiest work around.

●

The Elements of Comfort

You may not remember the precise time and place, but you've probably experienced — sometime in your life — a moment of nearly perfect physical comfort. The air temperature would have been about 70°F, let's say, and the humidity down around 50 percent. There probably wasn't any breeze or draft — at least not enough to notice. And the air, washed clean with rain or scrubbed with man-made filters, was momentarily free of bus fumes and cooking odors.

We tend to forget such moments, though they are pleasurable and rare. Instead, we remember those challenging moments when the summer sun turned us into frazzled balls of sweat or the winter wind made us shiver like leaves.

What we really want, in the privacy of our homes, is full-time, maximized comfort: that narrow range of conditions that enables us to forget about our bodies entirely.

Under normal conditions our bodies are maintained within a few tenths of a degree of 98.6°F. If we lose body heat too quickly — on a cold winter morning, for example — the body tries to cut heat losses by restricting the amount of blood flowing to the skin, hands, and feet. At the same time, we start to shiver, an involuntary muscular reflex that helps us keep warm.

To avoid having to shiver, we might add some body insulation (put on a sweater), or generate more body heat (exercise), or find a source of radiant heat where there aren't any convective currents (sit in front of a sunny, draft-free window), or increase the air temperature and relative humidity (crank the thermostat up on the furnace and turn on the humidifier).

If, on the other hand, it's one of those scorching afternoons in July and we aren't able to shed heat fast enough to stay comfortable, our bodies' natural cooling mechanisms kick in.

First, the blood vessels near the surface of the skin expand, producing a radiator-like effect that enables us to release more heat — via convection — into the air. If that's not enough, we start to sweat — the original evaporative cooling system.

To avoid having to sweat, we might strip away some of our insulation (peel down to shorts or swimsuits), or generate less metabolic heat (take a siesta), or find a cool radiant environment with lots

One of our primary goals in Santa Fe was to buffer the Ashers' home against summertime heat. Thick adobe walls, combined with sun-blocking overhangs and carefully positioned shade trees, help keep the house cool without placing excessive demands on the air conditioner.

of convective current (retreat to a shady porch with a fan nearby), or lower the air temperature and relative humidity (turn on the air conditioner).

The whole history of heating, ventilation, and air conditioning has been to alter the environment around us so that our bodies won't have to work as hard. To the extent that HVAC systems succeed in this, at a price we can afford, we become masters of our home environment. To the extent they fail, or tax us with heavy expense, we become prisoners in our own dwellings.

Some years back, I paid a wintertime visit to an acquaintance in Cambridge, just across the Charles River from Boston. Outside and in, my friend's three-story Victorian *looked* to be tidy and well maintained. But down in its basement and concealed behind its walls a very different picture lurked.

I found my friend hard at work in his first-floor study, hunched over his computer. He wore a wool stocking cap snugged down around his ears and a thick, fur-lined parka with a San Francisco 49ers' logo emblazoned across the back. He had cut the fingertips off his gloves, an age-old compromise between warmth and dexterity. Over by the window sat the telltale radiator, cold and silent.

"Man, it's kinda chilly in here," I said, with true understatement.

His breath frosted hard out over the keyboard. "The heating system's a *beast*," he said. "And an *expensive* one at that. So I just shut it off."

My friend, like a lot of other people I've met over the years, was a prisoner in his own house. It wasn't from a lack of money that he suffered (underneath his parka that day he wore a Brooks Brothers suit) but from a lack of information and a plan to make things better. Though studies show that we spend 90 percent of our time indoors, most of us don't understand all the variables that affect our comfort and health, or how we might change things to control them better.

When our homes turn chilly, drafty, or damp in the winter, or hot and humid in the summer, we generally point an angry finger at the boiler, furnace, heat pump, or air conditioner — that is, the mechanical system responsible for heating or cooling the air. But air temperature, as we'll soon see, is only one of many factors that influence comfort. And in many cases, it's not even the most important.

Thermal Comfort

By my count, there are six different factors that influence thermal comfort. These are air temperature, air velocity, relative humidity, radiant environment, your activity level, and the insulating value of your clothes.

Air temperature is by far the most common measure of comfort, and the one that's most widely understood. When the air temperature rises or falls, it affects how much heat our bodies give up via conduction, convection, and evaporation.

Conduction, as shown in the drawing, is the transfer of heat through a single material or from one material to another where their surfaces touch. The rate at which heat is lost (or gained) depends on the nature of the materials — in this case, skin and air — and the difference in their temperatures. Consider how your bare cheek — at about 98.6°F — feels on a winter morning when the air up against it is only 10°F. That big spread in temperature, with no scarf to insulate you, produces heavy conductive losses and big-league discomfort.

You'll feel even more uncomfortable if there's a wind sweeping across your cheek, producing convective heat losses at the same time. This is what the weather forecaster means when she warns us about "wind chill." Simply put, convective heat losses occur as cool air moves along a surface — in this case, your cheek — absorbing heat.

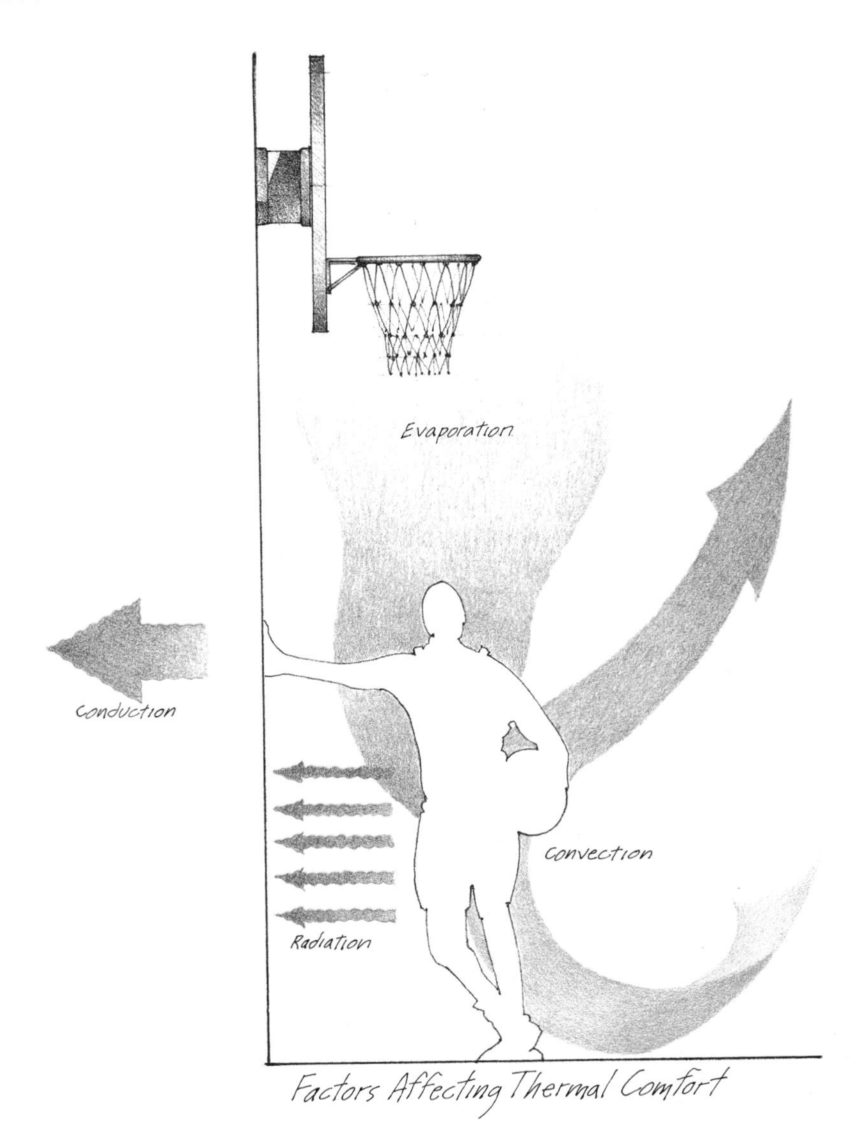

Factors Affecting Thermal Comfort

Though you might not be aware of it, there are also winds *inside* your house, which can have a powerful effect on comfort. Streams of air leaking through walls, windows, roof, and floor — known in the weatherization trade as "infiltration" — are a common source of indoor drafts, especially in older homes. Another source of indoor air movement are those natural convective currents created as warm air rises and cold air falls, setting up a cyclical flow inside each room. Finally, there are drafts created by forced air heating systems, air conditioners, and fans.

These interior winds are a frequent source of wintertime discomfort. Even in the summer, when moving air can be a godsend to a sweltering body, most people say they would prefer it if they could somehow feel comfortable *without* the draft. I'll have more to say about air movement in upcoming chapters on weatherization, heating, and ventilation.

Last but not least, air temperature has a powerful effect on how much heat we're able to shed by evaporation. Within certain temperature and humidity ranges, our sweat glands can provide over half a gallon of water per hour. As air passes over the skin, and that sweat evaporates, heat is drawn out of the body.

Here we begin to see how temperature and relative humidity work hand in hand in determining comfort. As humidity rises, we're less and less able to shed heat through sweating, which is why people in hot-but-dry Phoenix don't complain about the summer weather as much as they do in warm-and-muggy Washington, D.C.

Relative Humidity

A whole slew of household comfort and health problems stem from having either too much or too little humidity in the air.

When the relative humidity indoors starts to exceed 50 percent on a continuing basis, molds, mildews, bacteria, and other "bio-nasties" start to proliferate in the carpets, on the walls and ceiling, and — most aggressively, I've found — inside your favorite sneakers. At the same time, depending on the air temperatures outside and in, your windows and walls may start to sweat. Worse yet, moisture may condense inside the wall and roof framing, which sets the stage for rot.

As far as thermal comfort is concerned, a higher relative humidity is usually a plus in the winter — because you lose less body heat through evaporation — and a negative in the summer, when you want all the evaporative cooling you can get. To put it another way, you could probably save money on your wintertime fuel bill by adding humidity to your indoor air, since it would enable you to feel comfortable at a lower set point on your thermostat. Inversely, by dehumidifying indoor air in the summer, you could shave money off your air conditioning bill and sacrifice nothing in comfort.

The secret here is to not exceed those balance points where too much or too little humidity begins to cause problems. At relative humidities under 15 percent, for example, your throat and nasal passages start to feel like Death Valley, furniture joints may come unglued, book bindings crack, paint may not hold to some surfaces, wood floors shrink, and static electricity becomes a nuisance.

Later on, when we discuss ventilation and cooling, I'll give you some pointers on how to raise or lower humidity levels in your home.

Radiant Environment

The radiant temperature of the things around you — walls, floor, ceiling, furniture, plants, even the family cat — affects how comfortable you feel in your home. Since most surrounding surfaces are cooler than the human body, and the laws of thermodynamics tell us that heat always radiates from warm surfaces to cold, we're usually losing radiant heat to the objects around us. This is a comfort plus in the summertime, when we need to shed heat, but a minus in the winter.

In diagnosing comfort problems in the home, it's important to remember that air temperature is irrelevant in determining your home's radiant environment. The air temperature in your living room in mid-February may be in a very comfortable range, but if your body is losing lots of radiant heat to a cold wall or a big bare stretch of window, you're going to feel chilly just the same.

One of the most effective ways to improve your home's *mean radiant temperature* is to insulate floors, walls, and ceiling, which raises their radiant temperature in the winter and lowers it during the summer, making everyone more comfortable all year round. If adding insulation isn't practical, rugs and wall hangings are at least a step in the right direction.

Equipping windows with shades or curtains on the inside and sun screens or awnings on the outside is another good way to take control of your radiant environment. Or, if it makes sense to replace the windows altogether, you can take advantage of new glazing technologies that will both enhance your comfort and cut your energy costs.

Over the last few years low-emissivity, or Low-E, glass has become the standard in new windows because it makes homeowners more comfortable even as it saves them money. (Emissivity refers to a material's ability to give off, or radiate, heat in the form of long-wave infrared energy.) During the winter the special Low-E coating on the glass reflects radiant heat back into the house rather than letting it escape into the great outdoors. In the summer, the coating works in reverse, deflecting incoming heat *away* from the glass.

Personal Factors

The fifth and sixth elements that influence thermal comfort — activity level and dress — can vary a lot from one person to the next. People come in every size, shape, and age imaginable, each with a different metabolic rate and distinctive mode of dress. A middle-aged couch potato who has Slim Jims and root beer for breakfast is apt to have very different comfort requirements than does a twenty-year-old vegetarian who runs aerobics classes in the den. Neither of them will agree with Great-aunt Beth, who weighs 75 pounds and was born sometime before Woodrow Wilson took office.

The number of people who are going to live in a house, their level of activity, and the way they dress are important considerations in designing a good HVAC system. When you consider the fact that a person who's working or playing hard generates 600 watts of heat — not to mention the moisture, carbon dioxide, and odor he throws off — you begin to see how much difference people can make. Put a couple youngsters together doing the lambada and you've suddenly got the thermal output of a small electric heater.

Though you may not give it too much thought, choosing the clothing you put on in the morning is probably the most important comfort decision you'll make all day. If you live in Moose Jaw, Minnesota, and make a habit of slothing around the house in shorts and a tank top in the middle of January, you should share that information with your HVAC contractor right from the start. In other words, overdressing or underdressing for the climatic conditions of the day will make your HVAC systems work that much harder and swell your energy bills proportionately.

With that in mind, I advise you to dress smart. In the summer, choose light-colored, open-weave clothing that admits lots of air and promotes evaporation. As a general rule, the higher the proportion of cotton to synthetic fiber content, the cooler the material will be. Tightly woven garments in dark colors are better for the winter — the more layers you use, the more insulating power your clothes will have.

It's important to remember that you can be comfortable in your home with many different combinations of air temperature, air velocity, relative humidity, radiant heat environment, and dress.

A *decrease* in air temperature, for example, can be offset by an *increase* in the mean radiant temperature. Or vice versa. Similarly, people won't mind a lower air temperature in the winter if the velocity of the air is reduced or the relative humidity increased.

When I think about all the comfort variables that need to be managed in designing a good HVAC system, it always reminds me of that show where comedian Steve Martin tries to juggle cats.

The Broader View

Most discussions of comfort would stop right here, with the six aspects of *thermal* comfort we've covered. But I think it's important that builders and HVAC contractors — and the homeowners they serve — take the broadest possible view of comfort. This has to include:

Air quality

It seems obvious to me that you can't be comfortable in your home if you can't draw a lungful of clean air. We sometimes think people who live downwind of a dirty refinery or across the road from a slaughterhouse are the only ones suffering from air pollution. But the truth is, thousands of homes have pollution and moisture problems that originate *inside* the house.

On the one hand, these pollutants can be as harmless as cooking smoke and bathroom odors — a mere nuisance. On the other, if radon gas, carbon monoxide, or formaldehyde are present,

the situation turns deadly. We'll talk more about this in the sections on weatherization and ventilation.

Noise

My sympathies are with you if you happen to live near a busy freeway or under the flight path to O'Hare International. I don't plan to delve into soundproofing construction techniques here, except to tell you that designs do exist for sound-resistant stud walls and there are such things as acoustically rated windows. (If you can't figure a way to keep the noise out, you might consider buying one of those gadgets that sits on your bedstand, generating the tranquil sounds of ocean surf.)

My main concern here is going to be with noise coming from *inside* the house, generated by noisy HVAC gear.

Light

If you were to carry energy conservation to its extreme, you might want to build a house without any windows at all. No more conductive losses through the glass that way. And no air leakage around the sills and frames.

Fortunately, we're smart enough to know that natural light is a fundamental ingredient to human health. Medical researchers have discovered that people who don't get enough sunlight can develop seasonal affective disorder (SAD), which brings on anxiety, irritability, and depression. Even people who don't suffer from SAD know what it's like to have the "blues" or "cabin fever" after suffering through a week of cloudy weather or being shut up indoors too long.

Apart from its role in sustaining our physical and mental health, natural light has economic advantages too. For one thing, it reduces the need for artificial lighting, which shaves dollars off the electric bill. And as we'll see in an upcoming discussion on solar energy, sunlight can be an important source of free heat.

Peace of mind

Being comfortable in your home also means resting easy at night with the decisions you've made and their repercussions.

Some of the issues that may affect your peace of mind are fairly easy to quantify and put to rest. For example: Are you comfortable with the price of the new HVAC system and the debt you're going to have to take on to pay for it? Are you satisfied that its operating costs aren't going to eat you alive?

Other answers tend to be more elusive: Are you confident that your contractor is willing and able to back you up with good, long-term service? Are you comfortable with the environmental and economic impact this system is going to have locally? Nationally? Globally?

Consuming fuel and water in the home and discharging combustion gases and sewage always brings with them a moral cost, a political cost, a national security cost, and an environmental cost. I was flabbergasted by the results of a recent poll conducted by the Alliance to Save Energy in which 73 percent of the respondents said they would be willing to pay 15–30 percent more money for the fuel they use if the increase were used to fight global climate change!

A Whole-House Approach to Comfort

The key to enjoying comfort, health, and economy in your home is to look at the building structure and mechanical systems holistically, that is, to see the interconnectedness and importance of all the various parts.

Savvy HVAC contractors know that it is impossible to tinker with one part of the comfort equation without affecting some other part. Beef up the home's insulation and suddenly you don't need such a large furnace; change the old atmospheric furnace for a smaller, closed-combustion model and you alter the home's ventilation rate; change the ventilation rate and you've probably affected the air quality.

Following the lead of manufacturers like Amana, Carrier, Honeywell, and Lennox, some heating and cooling contractors are beginning to adopt this whole-house approach to comfort. They understand that smart controls, efficient distribution, and the proper amount of ventilation, air filtering, and humidification are every bit as important to household comfort as the central heating and cooling hardware. Some are broadening their horizons to include insulation and weathertightening work as well, or at least to understand the critical relationship between the home's envelope, comprising walls, windows, doors, and roof, and its HVAC systems.

"I used to think of myself as a 'weatherization' specialist," says Don Jones, of the Housing Resource Center in Cleveland. "I didn't want to be bothered learning about the furnace or the air conditioner, because those were somebody else's job. But this division of the trades, where everybody has his own little field of expertise, ends up creating major comfort problems for the homeowner. Because nobody sees the big picture."

Don tells me that he thinks of himself nowadays as a "recovering specialist" and that he's working hard to become a "good generalist" who understands how the different components all fit together.

•

Who Does What?

Working on the "This Old House" project in New Orleans reminded us once again how important it is to have a reliable and experienced foreman on the job. Master carpenter Norm Abram, who serves as general contractor on our projects, suffered a death in his family during the project and had to return to New England for the funeral.

Everyone on the crew was amazed, and considerably humbled, at how little we got done without him. Decisions that would have been second nature to Norm now had to be mulled and calculated, and sometimes reversed. Work slowed to a crawl.

That's because Norm is a lot more than a TV personality, or even a master carpenter. He has that special combination of skills it takes to oversee a complex job and carry it through to completion, which means he's part juggler, part timekeeper, part quality-control inspector, and part visionary. It's true, we learned in New Orleans, that you don't fully appreciate a talent like his until it's missing.

I sing Norm's praises here not to flatter him, or because he's a friend, but to emphasize the importance of professional leadership on the job and to encourage you to shop for contractors with similar qualities, namely: experience, honesty, friendliness, and reliability.

Investing Your Own Labor

Once you've decided that you need to make improvements in your home's HVAC systems, two big questions arise:

First: How much of the work, if any, do you want to do yourself?

Second: How do you choose a contractor?

Almost all of the homeowners appearing on "This Old House" have pitched in to help with the work in one fashion or another: demolishing walls, stuffing insulation, hanging wallboard, painting, cleaning up the job site, or landscaping. While I like the idea of homeowners' investing what I call "sweat equity" in their projects, it's important to draw some practical lines between what happens in real life and what makes good television.

One of the things that makes "This Old House" so enjoyable for us to produce, and for you to watch, is that we go out of our way

Top: One of the reasons that Norm Abram (left) is such an able project leader is that he makes it his business to understand the subcontractors' specialties and to oversee their work. Here he examines details on a ridge vent construction with roofing contractor Mike Mullane.

Bottom: Brian and Jan Iago, owners of our Lexington, Massachusetts, project, took the time to understand the tradeoffs involved in selecting HVAC equipment so they could make informed decisions.

to get the homeowners involved in the work, presenting as much of their lifestyle and personality as we can.

In real life, contractors are much less thrilled about the homeowner's picking up a hammer or wrench. Some will refuse to allow it at all, partly because of questions of liability, partly because it's so tricky to assign a monetary value to the homeowner's work. Most contractors start with the basic assumption — and oftentimes they're right — that the homeowner will only slow the work and compromise its quality.

At my family's firm, Trethewey Brothers, we have a little plaque up on the wall that, like all jokes, contains a serious element of truth:

Rates Per Hour	
Standard rate	$40
If the homeowner helps	$50
If the homeowner gives advice	$60
If the homeowner tells plumbing jokes	$70

I should also point out that the opportunities for sweat equity in HVAC work are far fewer than in general carpentry and face-lifting projects. Cutting balusters for a deck railing or hanging drywall is a world apart from installing a new condensing furnace or boiler. In point of fact, it takes three solid years of on-the-job training before a plumbing and heating contractor's apprentice is worth a hoot. How then can a homeowner expect to achieve miracles in the span of a weekend?

All this is not meant to discourage you from doing some of the work yourself. In fact there *are* opportunities — provided your contractor is willing — to help out with insulation, caulking, and other weatherization measures, all of which are fundamental to household comfort. And there may be opportunities in the finishing stages of the work and cleanup. We'll be detailing some of these *do-able* do-it-yourself projects in chapters to come.

The golden rule here is never to overestimate your own skills or the amount of time and energy that you can *realistically* spend on the project. And be prepared for a mishap or two along the way — Murphy's Law exacts its toll from amateurs and pros alike, usually when you least expect it.

One final word about sweat equity. In addition to having the appropriate skills, and the stick-to-itiveness to see the work through, you must have the right tools and the know-how to use them safely.

Perhaps some of you remember the amazing electric dolly we demonstrated on the show a few years back. All you have to do is edge it in underneath the object you intend to move — a boiler, say, or a refrigerator — and use its battery-driven wheels and hoisting mechanism to do all the work. Take it from me: renting this kind of labor-saving device is much preferred over hernia repair.

No matter what the job, be sure to find out ahead of time what kind of tools you're going to need to do it right, then figure out a way to buy, borrow, or rent them. Without the right tools at hand, a simple project can take forever.

Once you've got the right tools, you have to know how to use them safely, which is sometimes easier said than done, even for a professional.

Safety always, invariably, indisputably comes first. If you don't know how to use the tool, find someone who can teach you. Or run some carefully controlled tests — with the instruction manual glued to your elbow — until you're sure you've got the hang of it.

People who wade into a project with inadequate skills and/or improper tools are actually scheduling themselves a meeting with their insurance agent. What does it gain you, after all, in terms of pride or dollars and cents, if you finish the job yourself but end up losing a thumb in the process?

Finding the Right Contractor

In real life, homeowners don't have the luxury of knowing that Steve Thomas, Norm Abram, Tom Silva, and — why be modest? — Richard Trethewey are going to handle the job for them. In real life the yellow pages make no distinction at all between good, honest contractors and incompetent hucksters.

It goes without saying that you can't judge a contractor by the size of his firm or the dimensions of his ad. Some HVAC and plumbing companies have grown so large that they've franchised their operations over whole states or entire regions, using dozens of trucks and hundreds of employees. On the other end of the spectrum are the thousands of mom-and-pop firms that work out of a house or small office with a single set of tools and a five-year-old pickup. In the middle ground are medium-sized firms like Trethewey Brothers, which was founded by my great-grandfather in 1902.

While size and appearance aren't necessarily indicators of quality in a contracting firm, I think longevity is. A business does not survive two, three, or four generations by doing slipshod work and ignoring its customers' service needs.

Having said that, I'll quickly add that I've met some bright,

While "appearance" is only one criterion among many, it's smart to choose a contractor who is professionally dressed and well equipped.

young, first-generation contractors who impressed me very much. So don't use the age of the firm as your *only* criterion.

Regardless of whether it's an old-line firm or first generation, the company should be happy to give you a list of customer references. Don't bother to call the first name on the list, since it's usually the owner's brother-in-law, but do follow up on some of the others.

Your own neighbors and relatives can also be good sources for leads. There's no stronger testimony than the word of a friend, assuring you that the contractor performed his job well, on time, amiably, and within budget. You may want to double-check the candidate's reputation with your local utility and Better Business Bureau. And it's fundamental, of course, that the contractor be properly licensed and insured.

Now, as never before, it's essential that HVAC contractors stay well informed. This is no easy task considering the swift and radical changes that high technology has brought over the past ten years.

The use of microprocessors, variable-speed motors, special metals and engineered plastics, and a whole new generation of compressors and heat exchangers has given us heating and cooling machines that are — impossible as it seems — almost 100 percent efficient. Moreover, modern-day furnaces, boilers, and heat pumps are generally quieter and more durable than their ancestors, and when properly installed and maintained will provide homeowners with a superior measure of comfort.

Problem is, all these attributes were bought at the expense of simplicity. Technology, in this case, has outpaced most homeowners' ability to understand exactly what it is they're buying. Many heating and cooling contractors are similarly perplexed (though loath to admit it) and need to go back to school before they attempt to install or service state-of-the-art equipment.

One of the nice fringe benefits of working with "This Old House" is that I get information and product samples — unsolicited — from virtually every maker of plumbing, heating, and cooling gear in the world. (And you think you've got junk mail problems!)

But the average contractor doesn't have this luxury. He has to work hard to stay smart — not from year to year, as it used to be, but quite literally from day to day.

In this regard I think it's worthwhile for homeowners to choose contractors who are active members in their respective trade associations, because those associations provide reliable information and schooling. Trade associations also help develop and publicize technical standards, and ask their members to adhere to a certain code of ethics.

The National Association of Plumbing, Heating and Cooling Contractors, the Mechanical Contractors Association, the Sheet Metal Contractors Association of North America, and the Air Conditioning Contractors of America are four of the more prominent trade associations. Obviously, membership in these is no guarantee that you're not dealing with an incompetent or a huckster. It's only one criterion among many.

Be sure to ask your candidate contractor if he's been to school and/or had plenty of experience on the particular system you're considering. The newer the product, the more important this becomes. One of the qualities that distinguishes good manufacturers of heating and cooling gear is that they routinely call their distributors and dealers back to the factory for training. When you check the contractor's references, try to find a customer who has purchased a system similar to the one you're considering.

The Telltale Marks of Talent

There's an old joke in the trade that I never tire of telling: Contractors, they say, are perfectly willing to try something new . . . as long as their father did it before them.

It's true, I think, that the construction trades in general, and

HVAC contractors in particular, have been pretty conservative about accepting new products. Perhaps leery is a better word.

Partly this is due to the fact that there are some manufacturers out there with a we're-in-a-hurry-to-get-this-baby-out-the-door attitude, who have a nasty habit of testing their products on the contractor's back. With that in view, a healthy amount of skepticism and a little wait-and-see are important to a contractor's survival.

However, an experienced and well-schooled contractor should have no qualms at all about discussing new types of systems and components with you, for the very simple reason that he understands how they work, knows their pros and cons, and has confidence in his own ability to install and service them.

What's more, a quality contractor will understand the complexity and interrelatedness of the comfort issues we talked about in chapter 2. He'll want to come out to your house and do a thorough inspection. (Beware of contractors who try to solve your problem over the phone or make estimates without coming out to your house.) Good contractors will want to sit down with you and discuss your comfort goals — or problems — from a whole-house point of view, then review a range of options, weighing the cost and benefits of each.

I've learned over the years that homeowners rarely have a true fix on their comfort problems. They understand the symptoms all right — it's the causes and cures that elude them. What you perceive as a "furnace problem" may in fact be traceable to poor insulation. Or inadequate controls. Or a leaky distribution system.

A smart and honest contractor can make that deeper diagnosis for you and get you the most for your money. That's why I always tell people: Buy the contractor first, then the hardware.

If you can't find a contractor who will give you that kind of whole-house evaluation and some honest advice, I suggest you hire an independent energy auditor to do the diagnosis, then deal with the contractor on a separate basis.

Another telltale mark of a good contractor is that he won't shy away from discussing money. In fact he'll probably insist on talking about money right up front. That's because he knows that homeowners' dreams are always bigger than their budgets. And that the only way a homeowner can make smart choices is by knowing early on and accurately how much things are going to cost.

Finally, a quality contractor should be willing and able to suggest money-saving ideas. "Why invest $6,000 in central air conditioning?" he might ask. "In this climate you only have forty days' worth of real summer to contend with. If we only air condition a few key rooms, you can keep half that money in the bank."

Cabinetmaker Glenn Berger (right) shows me his plans for the house in Wayland, Massachusetts, so that I can make sure the radiant floor-heating system won't be obstructed by the cabinets. I always check in with the other contractors early in the project so that we can avoid headaches later on.

The Value of Friendliness

Having a contractor who's personable and easy to work with is more than a mere pleasantry — it's indispensable to getting a quality job.

To begin with, there's a process of investigation — oftentimes prolonged and personal — that has to take place so that the contractor can properly do his job.

If he's worth his salt, the contractor will ask for details about your lifestyle and scheduling. How do you actually use the rooms in your house? Are you home during the daytime, or off at work? Could we take a look at last winter's fuel bills? Where is the draftiest spot in the house? The darkest? The coldest? Does Aunt Ethel, who's seventy-three, or son John, who's fourteen, have a different point of view on all this?

If you've chosen your contractor well, he'll have a knack for asking insightful questions and be attentive when you talk.

Be frank with your answers. It's his job to extrapolate what you tell him into a plan that can both address your comfort problems and accommodate your budget.

It's also important that *you* listen well. When your contractor offers advice, he's sharing the benefits of his long and worthwhile experience. And that, primarily, is what you're paying him for.

After the process of discovery is over, you're going to have to live with your contractor for a while. Sometimes it's a matter of days. Sometimes weeks, or even months. So it's important that you and your contractor get off on a friendly footing.

Whenever I'm starting a new project, I always look my clients straight in the eye and say: "Because of the nature of remodeling work, you're probably going to hate me at some point in this project. But by the time it's over, you'll love me again." For the most part, that's held true.

There's another good reason why your contractor ought to have an agreeable — or at least tolerable — personality: He needs to get along with other professionals on the job. The more complex the project and the more players you have, the more important this becomes. Some jobs end up resembling a veritable alphabet soup of professions, with AIAs (architects), PEs (professional engineers), CRs (certified remodelers), CKDs (certified kitchen designers), CBDs (certified bath designers), ASIDs (American Society of Interior Designers), and so on and so forth all cluttering up the site.

If the heating and cooling contractor can't get along with the rest of the crew, and everyone starts reading from a different script, you're left with the proverbial Chinese fire drill. It's cooperation and a common sense of purpose between the professions and trades that produces excellence.

One way to verify a contractor's disposition on the job, as well as his talents, is to talk with other builders and subs who have worked with him before.

Be sure to remember your own responsibility to be courteous. You'll get better results that way. By taking the time and trouble to hire smart in the first place, you won't have to worry about playing spy or fussbudget later on. Trust will take care of it.

Looking beyond the Price

You may want to interview several contractors before you make your final choice. Invite them into your home. Visit them in their offices. If you can, talk with the person who will actually do the work.

If you're lucky enough to find more than one contractor you feel confident with, get competitive bids on like systems. Compare values on the warranty and service contract as well as the hardware.

I urge you *not* to make price your sole criterion. There are thousands of HVAC contractors out there who specialize in whacking in the cheapest equipment they can find in the fewest hours possible. But you're only going to get what you pay for. If one bidding contractor is radically lower than the other one, there's a good chance that he won't finish the job properly and/or hasn't left any money in the deal for service.

Both you and your contractor have an obligation to understand that heating and cooling systems have a very long life, affecting comfort and economy long after you've forgotten about first cost. When you hammer the contractor for the cheapest, cheapest, cheapest, you share the responsibility for the discomfort and heavy operating expenses you suffer over the life of the system.

Cementing the Deal

Some contracting jobs still go forward like they did it in the old days — on the shake of a hand. But in these litigious times I think it's advisable to have a contract.

Most contractors will have a standard form that can be amended to fit almost any project. If you like, you can have your lawyer look it over or buy an hour or two of time from a firm that specializes in construction law.

The contract need not be long and formal, but should be balanced in a way that protects one and all. It should:

- certify that the contractor is properly licensed and insured;
- outline project tasks and assign responsibility for each;
- describe materials;
- specify those items that are *not* included in the job;
- set a timetable, including start and finish dates;
- establish a clearly defined payment schedule, including the amount of the retainer and terms for paying the balance;
- define what portion of the work, if any, the homeowner will complete, and how that work is to be valued in the overall cost of the project (the contractor may want you to sign a liability or insurance waiver);
- spell out warranties (ask to see warranty documents on major components and determine if the contractor covers those items disclaimed by the manufacturer);
- schedule a periodic evaluation of progress and an end-of-project review;
- provide a process by which the contract can be changed — this is usually done in writing, through a formal "change order";
- include an arbitration clause to settle disputes;
- guarantee the right of recision (72 hours to change your mind).

Try to do all your thinking and equivocating during the contract and planning phases of the project, when changes are easy and cheap. As we've pointed out on the show many a time, changes get very expensive once construction is under way.

●

First Things First

Along with the rest of the crew at "This Old House," I've been preaching the principle of "conservation first" for a good many years now.

Trying to heat and cool a leaky, poorly insulated house is like trying to draw a bath with no plug in the drain. No matter how much hot water you pump in, the tub never fills. If people really understood how many of their energy dollars went straight down the drain, they'd make it their business to learn more about plugs.

We've put this advice into action on the show time and time again, always with good results. In fact, the very first "This Old House" — a circa 1860 Victorian in Dorchester, Massachusetts — was a textbook study in how to turn a cold, drafty, energy-gobbling monster into a comfortable and energy-efficient angel.

Those of you who were with us during that first season may recall the effort we put into caulking and weatherstripping. By the time the job was done, we must have gone through 100 tubes of caulk and half a mile's worth of weatherstripping.

Some of the windows — too far gone to save — were replaced with new units; others were simply reputtied. Storm windows and doors were added to shore up the primary windows. Where studs and joists were open and accessible, we laid in fiberglass batt insulation. Where they weren't, we had cellulose insulation blown in.

One of our guiding principles — then and now — is that tightening up a home's "envelope" and adding insulation are the two most cost-effective steps you can take toward affordable comfort. (The envelope refers collectively to a structure's walls, roof, windows, and doors.)

Even if you've been around the block on some of these measures before, I advise you to take a second look. Over time, caulk and weatherstripping can grow brittle or break away. New leaks can appear each season. Chances are that energy costs in your area have risen to the point where an additional layer of attic insulation would pay off. Or maybe you're planning an interior remodeling job or considering new siding — perfect opportunities to add more insulation inside or out.

Converting this old Victorian in Dorchester, Massachusetts, from an energy hog into an energy-efficient angel required every trick in the book.

Better than CDs

It's ironic, I think, when people invest their money in bank certificates of deposit or money market funds that pay 4 or 5 percent a year when an investment in home weatherization can often yield 10 or 12 percent *tax free.*

The problem, of course, is that the yield on weatherization is "invisible" in conventional terms — that is, it's realized in money that you *don't* have to spend. It's like a secret Swiss bank account, pumping money into your pocket year after year, with no interest or dividends to declare. And here's the clincher: You get a house that's a lot more comfortable to live in at the same time.

While it's hard to quantify the value of added comfort, the dollar return on your investment *can* be measured — in the form of lower gas, oil, and electric bills. In other words, the money you spend on airtightening and insulation will let you reduce the operating rate of your boiler, furnace, or heat pump during the winter and to use less air conditioning in summer. And when it comes time to buy new HVAC systems, you'll be able to specify smaller, less-expensive equipment. Should you decide to sell your house, you may even realize a capital gain, because smart buyers recognize the added value of an energy-efficient home.

Setting Priorities

Perhaps the biggest challenge in weatherization and HVAC work is determining exactly where to sink your money when you're working on a limited budget. One way to establish priorities is to call in a competent and objective energy auditor who can pinpoint the structure's weaknesses.

Home energy audits are easily arranged through your local utility, insulation contractors, HVAC contractors, state energy offices, universities, home inspectors, or Energy Rated Homes of America, a not-for-profit organization that's active in more than a dozen states.

Many utilities and heating oil companies will audit your home for free, or for a nominal fee, but the quality of their inspections and the value of the recommendations can vary widely.

Utilities that are strapped for peak generating capacity and want to put off building expensive new generating plants usually have a real vested interest in conserving energy in their customers' homes. Such companies can provide you with a serious, detailed energy audit, valuable site-specific recommendations, and referrals to competent contractors.

A surprising number of utilities now offer home weatherization loans on favorable terms. Customers of Austin Power & Light, for example, are eligible for five-year home energy loans at 3½ percent. Madison Gas & Electric, in Madison, Wisconsin, will loan its customers up to $25,000 — at below-market rates — to purchase anything from energy-efficient doors to high-tech furnaces. Other utilities pay cash rebates to homeowners and builders who weatherize or buy high-efficiency HVAC equipment and appliances. Some even give away energy-saving devices.

To illustrate how serious some utilities are about saving energy, consider the fact that Pacific Gas & Electric — with 7.7 million customers — plans to spend $2 billion on conservation during the 1990s. Consolidated Edison, New York City's main supplier, has an energy-conservation plan stretching well into the next century, *budgeted at $4 billion.*

Unfortunately, for every utility and fuel oil dealer that's serious about home energy audits and conservation, there are two or three others that are only going through the motions. If you have a capacity-rich utility that's only interested in mollifying its regulators, your "free home energy audit" may turn out to be a sham. Don't be surprised if you get a moonlighting science teacher who strolls quickly through your house, hands you a pre-prepared checklist of recommendations, and bids you a curt farewell.

Some weatherization and HVAC contractors provide good, objective audits and valuable recommendations, with fees ranging up to about $150. But most contractors arrive on the scene with a vested interest and may not have a real understanding of how the house's envelope and HVAC systems interact.

It comes as no surprise, I'm sure, when I tell you that insulation contractors are interested in selling insulation. And guess what? HVAC contractors have more than a passing interest in selling new heating and air-conditioning gear. The question is: Does the guy have enough integrity to give you the straight scoop on what you need even though it may not be his most profitable answer?

Furthermore, many weatherization and HVAC contractors don't understand how the various pieces of the house work together. As I mentioned before, I think this schism between the trades is going to disappear before too long, with very positive results.

For example, Amana, a leading manufacturer of heating and cooling systems, is busily training its Texas dealers — guys who have typically been hardware oriented — how to diagnose and treat a house as an integrated whole. The first thing they do for their customers is a free (serious) energy audit and blower door test, so that air leaks in the home's envelope can be precisely located and plugged.

If it turns out that the home's most pressing need is airtightening work or additional insulation, then that's what the Amana dealer will recommend and bid. If he finds that inadequate ventilation, leaky ductwork, or bad thermostats are at the root of the problem, the customer will hear about it. This is not to say that Amana is abandoning its core business — that is, selling and servicing heating and air-conditioning equipment — only that its dealers are being trained to address the full spectrum of problems that can affect household comfort and economy.

"I think it's unethical to drop a new high-efficiency heat pump or central air conditioner into these houses and tell people it's going to solve their comfort problems or cut their utility bills in half," says Tom Mooney, who oversees Amana's south Texas dealer network. "Unless you address the air-infiltration, insulation, and control and distribution problems first, you may as well brace yourself for callbacks and complaints."

Tools of the Trade

In the old days, auditors and weatherization contractors came to the job site with their caulk guns, a bag full of weatherstripping, and plenty of intuition. They had no way of knowing exactly where the leaks were or — once the job was done — how successful they'd been at sealing them. Likewise, HVAC contractors had to size heating and cooling equipment using ambiguous standards, because there was no way of really measuring how much hot or cool air would be lost through the home's walls and roof.

Then along came the blower door, a high-speed, calibrated fan that's mounted in a door-sized frame that adjusts to fit snuggly in the house's main doorway. When the fan is turned on, it pressurizes (blows into) or depressurizes (blows out of) the house, forcing air through cracks in the house's envelope. Workers go from room to room with smoke pencils, noting the places where air is snaking its way out of (or into) the house.

As caulking and weatherstripping work proceeds, the blower door is run periodically to measure progress. Once the big holes are sealed, it puts pressure on smaller leaks, which are in turn plugged. The idea is to tighten the house to the point where maximum comfort and energy-efficiency gains are realized, but not to the point where moisture or air-quality problems start to crop up.

Professional air-tightening crews usually roll the cost of the energy audit and blower door test into the final contract price, which can run from $.40 to $1 per square foot. A 2,000-square-foot house, for example, might run about $1,200. By the way, if the contractor promises to save you a certain percentage on your gas, oil, or electric bill, or makes other guarantees, be sure to get those promises down in writing.

Blower door tests can also help HVAC contractors estimate how many air changes per hour (ACH) occur in the house. If a house has 1 ACH, for example, it means that naturally occurring ventilation — through the walls, floor, and roof — will replace the total volume of indoor air with fresh outdoor air once every hour. A tightly built house might have .5 ACH, while a leaky one could have 2 ACH or more.

Those numbers are important to an HVAC contractor in precisely sizing heating and cooling gear and in deciding whether or not a house needs mechanical ventilation to maintain good air quality. A blower door test can also be used to measure and pinpoint the duct leaks in a forced air heating system.

Mooney is confident that the other big manufacturers of heating and cooling gear will soon follow Amana's lead.

"By the end of the decade, you're going to see HVAC contractors coming to the job site with their blower doors, performing a thorough energy audit, and addressing the thermal envelope and distribution problems first," he predicts.

Until that day arrives, there are other good places to find an audit. The American Society of Home Inspectors (ASHI) has more than a thousand practicing members around the country that specialize in doing objective inspections. They're often called in on real estate deals to assure a would-be buyer that his dream house isn't concealing radon problems, termites, or other nasty surprises.

A full ASHI inspection, which would include the home's foundation, walls, and roof, weathertightness and insulation, electrical systems, plumbing, and HVAC, may take up to three hours and cost about $250. A limited inspection, focusing on the thermal envelope and HVAC equipment alone, would cost about $100. In either case, you receive a written report on your home's energy and comfort status, and a list of recommendations you can trust.

ASHI includes many architects, engineers, and builders in

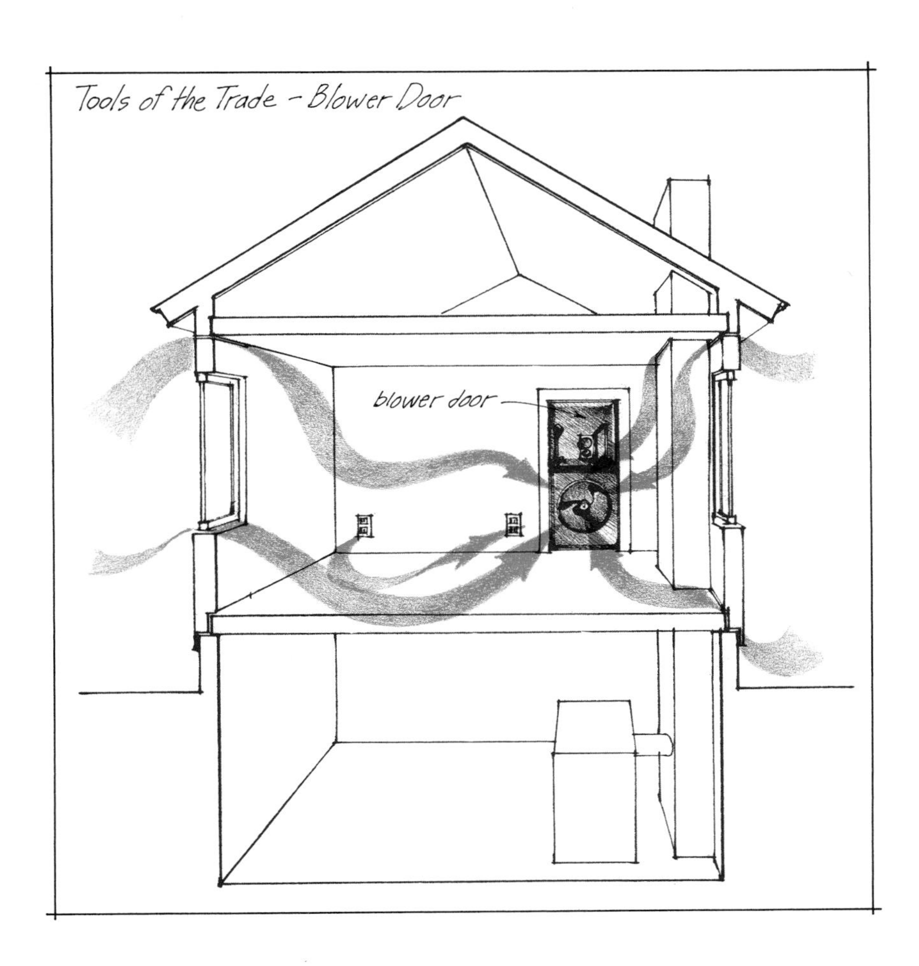

its ranks, who are required to pass three written examinations and to have at least 250 home inspections under their belt before they're admitted. Inspectors are also required to earn continuing education credits each year so that they're up to speed on the latest technologies. Best of all, ASHI's code of ethics prohibits inspectors from recommending contractors for repairs or from doing any of the repair work themselves, and forbids any collusion with real estate agents. That means they won't have any vested interests in the audit's results.

I'll warn you ahead of time, though, that there are a lot of new boys on the block when it comes to home inspections, and a fair number of them are incompetent, dishonest, or both. If you're considering an inspector who's not a member of ASHI, check out his credentials with extra care. Don't be deceived by inspectors who claim they're "certified" (by whom?) or "fully licensed" (there is no licensing process for home inspectors, though some states are considering it).

Another reputable source for home energy audits is Energy Rated Homes of America (ERHA), which completes each audit by assigning the home an overall energy rating. ERHA, a nonprofit organization based in Little Rock, Arkansas, is pushing hard for quality and consistency in its ratings, so that banks and mortgage companies will feel confident using the results as documentation for Energy Efficient Mortgages and home equity loans. In other words, the rating must convince the lender that more lenient terms on the loan are justified because the homeowner will have fewer energy costs to cover.

"The traditional approach in mortgage banking has been to add up the combined costs of the home's principal, interest, taxes, and insurance — called 'PITI' in the trade — and to weigh that sum against the applicant's income," says Ron Hughes, president of Energy Rated Homes of America. "Now, bankers are beginning to understand how important a home's energy costs are to that equation, and to extend a higher debt-to-income ratio for energy-efficient homes."

State-of-the-art inspections, such as those conducted by Energy Rated Homes of America, use blower door tests and sophisticated computer programs that include information on the local climate, fuel prices, and material and labor costs. Using special software, the auditor integrates these variables with the information he's collected on the house, generating a cash-flow analysis on a whole range of possible energy-conservation measures. Once the numbers are crunched, it's easy for a homeowner (and the homeowner's banker) to see which measures produce the biggest bang for the buck.

The case study presented here — courtesy of Energy Rated Homes of Vermont — shows how a homeowner in North Ferrisburgh, Vermont, was able to finance $12,000 in energy improvements on his house and enjoy a positive cash-flow position from day one. In fact, the net yearly savings to the owner, over and above the incremental cost on his mortgage, was $480.

While this example involves the purchase of a new house, the idea can work equally well for people who already own a house and want to upgrade its efficiency and comfort with a home equity loan. Who says there's no such thing as a free lunch? Actually, it's even better than that — it's like being *paid* to have lunch.

In addition to the twelve states that already have Energy Rated Homes of America programs, nineteen others are considering the plan. You can find out through your state energy office if it's available in your area. And while you're asking, ask them for names of banks and mortgage companies that are writing Energy Efficient Mortgages. One private company that has been a leader in this area is GMAC Mortgage Corp., which writes hundreds of millions in first mortgage and home equity loans every year.

Financing an Energy Overhaul

When Thomas and Julia Rood considered buying a 2,400-square-foot farmhouse in North Ferrisburgh, Vermont, they recognized that beneath its charm and beauty, the place was a bona fide energy hog. With insulation nowhere in sight and a boiler that dated back to the Hoover administration, the couple decided to have the house rated by Energy Rated Homes of Vermont (ERHV), a not-for-profit organization affiliated with the national Energy Rated Homes of America.

The raters checked out the old farmhouse and gave it a one-and-a-half-star rating, which put it somewhere between "Poor" and "Not So Hot." If nothing was done, they reported, the Roods could expect to pay $3,100 a year for oil and electricity.

As a result of the rating, Energy Rated Homes of Vermont recommended a complete package of improvements to bring the house up to a four-star rating. The measures included caulking, weather-stripping, and other steps to reduce air leakage, window repairs and upgrades, new attic and wall insulation, a new oil boiler with an indirect-fired hot water storage tank, and an automatic setback thermostat. Because the farmhouse was in such sad shape, the total cost of these improvements would come to a whopping $12,000. (A more typical cost to upgrade a more typical house would be $3,000 to $4,000, ERHV says.)

Using ERHV documentation at the bank, the Roods were able to finance the energy improvements as part of their mortgage. The upgrade added about $110 to their monthly mortgage payment. But their annual oil and electric bill shrank to only $1,260 a year, an annual savings of $1,840, or more than $150 a month. So the Roods were in a positive cash-flow position right from the start, saving $40 a month, or $480 a year. And enjoying the added comfort that the improvements provided as well.

The Rood House, Ferrisburgh, Vermont.

When the Roods put their house on the market a couple of years later, in a slow economy, it sold in just three days.

"The improvements really made a difference in selling our home," says Julia. "People love old houses, but they love them even more when they know they won't have to spend a fortune to keep warm."

Economics of a Free Lunch

	Star Rating	Additional Mortgage Payment	Energy Cost
Home in present condition	*½	$ 0	$255/month
Home after improvements	****	$110/month	$105/month
Difference		+$110/month	−$150/month

Net savings per month: $40
Net savings each year: $480
Net savings after 30 years: $31,891[1]
Net present value of savings after 30 years: $7,222[2]

1. Assuming a 5 percent annual fuel inflation rate but not taking into account the time value of money.
2. Assuming a 5 percent annual fuel inflation rate; 10 percent discount rate.

SOURCE: Energy Rated Homes of Vermont.

In what may be a preview of things to come in other states, the Iowa state legislature recently made home energy ratings mandatory for both new and existing homes. The California legislature is considering a similar law.

Tips for Do-It-Yourselfers

If you decide to forge ahead on your own, without having a professional energy audit done, common sense suggests a few precautions. Before you drain your bank account, make sure the money you plan to spend on energy conservation is proportionate to your bills. It doesn't make sense, for example, to invest $10,000 in energy conservation if your heating, cooling, and hot water bills are only running $800 a year. Even if the investment reduced your bills to nothing — an impossibility — it'd still take more than twelve years to recoup your money.

On the other hand, if your utility bills were $2,500 a year, and you spent $4,500 to achieve a 35 percent savings — a *real* possibility — the investment would pay you a tax-free annual return of 19 percent. After recovering your initial investment — in about five years — the annual savings would be an extra bonus.

In making these calculations, be sure to consider how long you're going to live in the house. Will you stay long enough to cash in on your investment and enjoy the extra comfort it provides? If you're in your retirement years or plan to move in a couple of years, investing for the long term may not make much sense.

As I mentioned before, it pays to put your money into simple measures first: caulking, weatherstripping, and insulation. If you're out shopping for a high-efficiency furnace and still haven't weatherstripped the front door or put an insulation wrap on your hot water heater, you need to sit down and rethink your priorities.

Homeowners who want to do their own weathertightening work should bear in mind that it's usually false economy to buy the cheapest products around. If you do, you'll probably end up having to redo the job somewhere down the line. For example, caulks made out of acrylic latex, polyurethane, and silicone are expensive compared to polyvinyl acetate and oil-based caulks, but they're true champions when it comes to longevity. Once you've factored in the labor costs of having to do the same job twice, or even three times, good quality materials turn out to be a bargain.

The same principle holds true for weatherstripping. Independent tests show that foam-type weatherstripping (urethane foam or silicone) survives temperature swings and wear and tear better than thermoplastics (polyvinyl chloride, polyethylene, polypropylene, and thermoplastic elastomers). The guidelines laid out on pages 46 and 47 can help you choose the right materials.

Do-it-yourselfers can get good results by imitating the strategies used by professional weatherization contractors, who usually start up in the attic and work their way down to the basement, crawl space, or slab.

Shopper's Guide: Caulk

Type	Approx. Retail Cost[1]	Lifespan[2] (years)	Comments
Oil-base	$1.00–$1.60	1–3	Won't adhere well to wet surfaces; discolors over tar or asphalt; paintable; cleans up with paint thinner; not recommended because of relatively short life.
Polyvinyl acetate (PVA)	$1.30–$1.60	1–3	Not recommended for outdoors or wet environments; can be used as general-purpose household adhesive; readily paintable; cleans up easily with water; relatively short life makes PVA unacceptable for most uses.
Styrene butadiene rubber (SBR)	$2.00–$2.25	3–10	Will adhere to damp surfaces and treated lumber; available in various colors; easy to work with; indoor application requires ventilation; takes latex paints; cleans up with paint thinner.
Butyl	$2.20–$2.90	4–15	Adheres well to masonry and metal, but not to damp surfaces; difficult to apply; long curing time before it can be painted; difficult to clean up, using paint thinner or naptha.
Acrylic latex	$1.50–$3.00	5–20+	Good adhesion to most materials, including damp surfaces; easy application; available in limited colors and paintable; easy clean-up with water; excellent general-purpose caulk.
Kraton base	$4.00–$7.00	5–15	Very good adhesion to almost every material; comes in clear or pigmented versions; shouldn't be tooled, because it shrinks to fit; scrubbable and paintable; cleans up with mineral spirits; rivals silicone in performance.
Polyurethane	$4.00–$9.00	5–15	Sticks well to masonry and most other surfaces (some may require primer), but not to damp surfaces; difficult to apply neatly; very tough and tear resistant; no shrinkage; comes in various colors; paintable; cleans up with paint thinner, acetone, or methyl ethyl ketone (MEK).
Silicone	$3.75–$6.00	20+	Good adhesion to most surfaces (some requiring primer), but not to concrete, stone, and some porous materials; won't shrink or harden with age; can be applied outdoors at low temperatures; requires care in preparing surfaces; may require special paint or not be paintable; cleans up with paint thinner, naptha, toluene, or xylene.

1. Per 11-oz. cartridge.
2. With exterior use.
SOURCE: Macklanburg-Duncan, Inc.; Madison Gas & Electric Co.; *Fine Homebuilding*; and Massachusetts Audubon Society.

The biggest problems — especially in older houses — tend to be bypass leaks, where hollow walls, closets, plumbing chases, ducts, and electric lines were left unsealed at the top and bottom, creating hidden chimneys up through the middle of the house. As cold air funnels up from the basement or crawl space, through the house, and into the attic, it can sweep enormous amounts of heat out of the adjoining spaces and create serious comfort problems.

Trethewey's Tips: On Caulking

Caulking a crack may look like child's play, but if you don't choose the right caulk for the job, and apply it correctly, you may as well use peanut butter.

Some caulks won't adhere well to certain surfaces (i.e., silicone to porous materials like brick and cedar). Others are designed for special applications, like caulking bathroom tiles, sealing vapor barriers, or chinking log cabins. So make sure the caulk you choose is compatible with the surfaces you're going to work on, and follow the manufacturer's instructions on how to clean those surfaces before you start. Otherwise, you won't get a good, lasting bond.

The manufacturer's label or printed instructions will also tell you how well the caulk works at different temperatures, its shrinkage characteristics, whether or not it's paintable, and what you'll need to clean it up. One of the things that makes acrylic latex caulk my personal favorite for a lot of household uses is that it cleans right up with water.

I wouldn't recommend bargain basement caulks (i.e., oil-based and polyvinyl acetate compounds) because they last only a year or two before they go kaput.

You'll get best results with your caulk gun if you cut the tip of the cartridge off at a 45-degree angle. Use a pushing motion as you work the gun along, holding the cut opening of the cartridge parallel to the joint. Make sure you inject enough caulk into the crack to form a good bond, without trapping air.

Master caulksmen always "tool" the joint after they've finished. Using a plastic spoon, butter knife, or professional caulking tool (but not your finger!), press the caulk down into the joint until it's left with a slightly concave surface. Not only does this leave the bead with a neat and even appearance, it also enables the caulk to flex more easily if the joint moves.

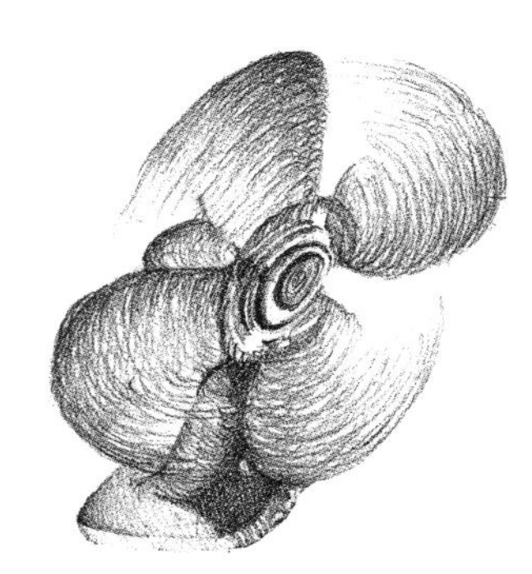

There are lots of different materials available for plugging cracks, gaps, and holes in your house. For small gaps, caulk alone will do the job. You can also use caulk for larger, deeper cracks — up to an inch wide, for instance — provided you fill the opening with backer rod first so that the caulk has a place to seat. Backer rod (also known as caulking cord, rope caulk, oakum, or crack filler) is sold in rolls at hardware stores and home centers.

Polyurethane foam sealant, which comes in 12- to 36-ounce cans, is a good (though costly) choice for voids up to about four inches. Since the foam expands after it's applied, you'll need a little practice before you can shoot in just the right amount. Use too much of the stuff and it comes globbing up out of the void. (No need to panic, though — just wait till the goop dries and trim it with a knife.)

For still larger holes, you can wedge in a piece of expanded polystyrene (beadboard) and caulk it carefully around the edges. If you want a little stronger material and better insulation, use extruded polystyrene (blueboard or pinkboard) instead. Neither type of polystyrene should be used around furnace stacks, chimneys, or heating pipes where the temperature can exceed 140°F — use metal flashing and silicone caulk instead.

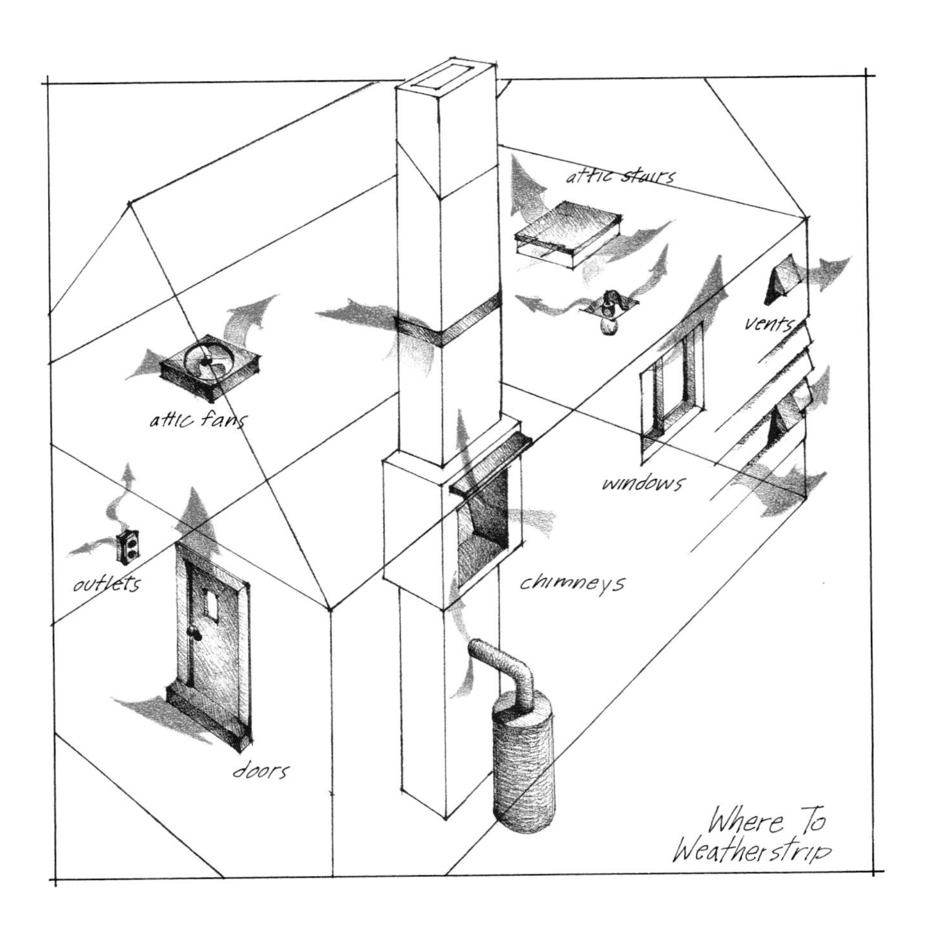

Unfaced fiberglass insulation is also a good choice for plugs, so long as it's wrapped in plastic or stuffed in a plastic bag. Without the plastic, the loose fiberglass won't do much to stop air leakage.

No matter what kind of plugging material you end up using for the larger holes, be sure to seal the edges tightly, using staples, caulk, and/or a high-quality tape that's suited for the job.

As you move down from the attic into the conditioned spaces of your home, the most common places to find leaks are around windows and doors, electrical outlets, plumbing penetrations, and baseboards. Recessed lighting fixtures, dropped ceilings, and unsealed attic and cellar hatches are also notorious culprits.

Some do-it-yourself weatherization work is so cheap and easy that it's always worthwhile. Putting foam gaskets behind all the electric outlets and light switches would fall into that category. So would putting sweeps and weatherstripping on exterior doors.

But short of caulking and weatherstripping *everything*, how are you to know — except through luck or intuition — which spots really need the work?

One way to put a little more science into the job is to rig yourself up a makeshift blower door, using an ordinary window fan and an incense stick. Start by placing the fan in an open window so that it blows air *out* of the room. Be sure to seal up the edges around the fan with rags or foam rubber and to close all the other windows and doors in the room, including the storms, before you start.

When you turn the fan on, it will depressurize the room, increasing the air flow through all its cracks, crannies, and crevices. By moving the smoking incense stick along the baseboards, around the windows, and over and across other suspect areas, and watching the trail of smoke, you'll be able to find currents of leaking air that would have otherwise gone undetected.

An attic fan can serve the same purpose, provided you bottle up the house before you start. If you don't have a fan, wait for a windy day and open up a single window or door on the downwind side of your house. This tends to pull a slight vacuum on the inside of the house, accentuating the leaks.

Windows of Opportunity

If your house falls into the "typical" category, you're probably losing about 25 percent of your heat through the windows. The first step toward cutting those losses and making yourself more comfortable is to replace any broken window panes and reputty worn-out seals. If the sash lock doesn't hold the upper and lower sashes snugly together when it's locked, spend a buck and a little time to replace it.

Next, check the seam where the window casing meets the wall — if there's a crack there, seal it with a paintable caulk.

Weatherstripping is one of those dull projects that can produce very exciting results. In fact, weatherstripping can save you up to $10 per window *every* winter and eliminate some very nasty drafts. It will also help reduce condensation on the inner surfaces of the window, which can damage wooden parts.

Spring strips, jamb-up gaskets, foam strips, and magnetic vinyl can all be used to good effect, but if you're like me — that is, born lazy — you'll want to choose the most durable types so that you won't have to redo the job anytime soon. (See our Shopper's Guide on pages 46–47.)

One way to tell if your old weatherstripping is shot is to lay a dollar bill (a *twenty* if you're rich) across the sill and close the window on top of it. If the bill slips out easily, it's time to put in new weatherstripping. You can use this same test on the meeting rail, where the upper and lower sashes come together.

So long as you buy the right piece for the right place, there's not much else to know about weatherstripping. With that in mind, I advise you to go ahead and read the package instructions while you're still in the store.

Another really effective and affordable way to upgrade old windows that have single-pane glass and/or worn-out sashes is to buy a sash replacement kit. Instead of tearing out and replacing the whole window — a very expensive proposition — only the upper and lower sashes are replaced. The window frame, including the sill, jambs, stops, and all of the trim, are retained.

Once the old sashes have been taken out of the frame, the sash pulleys and weights are removed, and the cavities filled with insulation. New jamb liners are then installed, which provide the tracks for the new sashes. Depending on the size and the quality of the replacement sashes you order, and whether or not you do the work yourself, a sash replacement kit will save you 50 to 75 percent compared to the costs of a new window.

Before you buy a sash replacement kit, make sure that the existing window frame really is sound, so that you don't end up having to redo the job a few years later. You can test the wood by probing it with an ice pick or screwdriver. The sill, particularly at the corners, is usually the first place to show rot.

Once you've rehabilitated your primary windows, follow the same steps with your storms: repair, caulk, and weatherstrip. If you don't have storm windows and doors, they'd probably make a smart investment, especially if your primary windows are older, single-pane units. The number-crunchers at the U.S. Department of Energy tell me that adding storm windows and doors can chop heating and cooling bills as much as 15 percent, provided the existing sashes are in good shape.

Permanent, combination screen-and-storm windows (triple-track) will cost you $50 to $110 per window, depending on the size and quality. Unless you do the installation yourself, contractor fees will add another $10 to $25 per window. In any case, it doesn't pay to scrimp on quality — spend enough to get good durable frames, tight seals, and weep holes to drain off excess moisture.

Another way to improve your windows is to install insulating quilts, which provide a blanket-like covering over the inside of the window. Close-woven styles with metalized linings and a tracking system that holds them snug to the casing can help seal out drafts and deliver R-values of 3 or better. (R-value refers to a material's ability to resist the transfer of heat. The higher the R-value, the more insulating power the material has. Single-pane glass, for example, has an R-value of about 1, while a 4-inch-thick blanket of fiberglass insulation provides an R-value of 13.)

Shopper's Guide: Weatherstripping

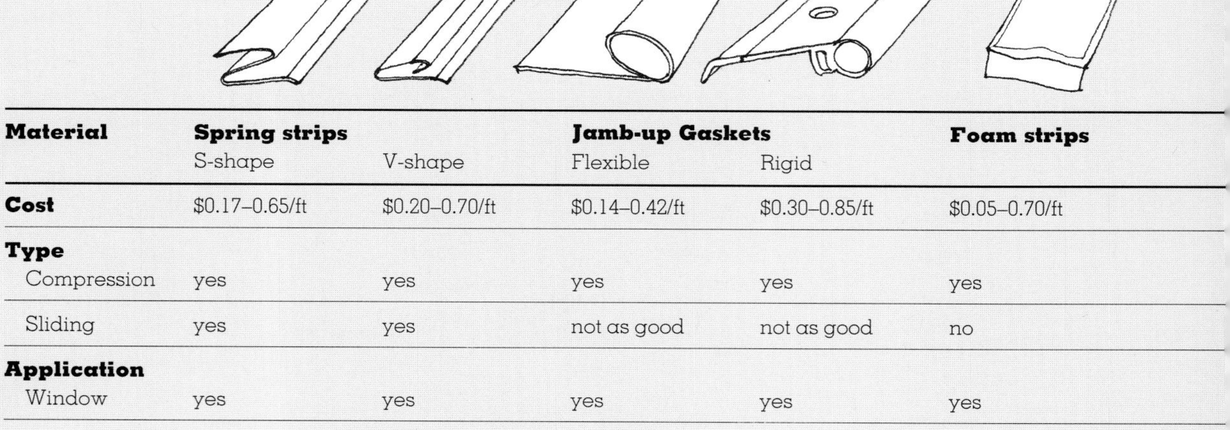

Material	Spring strips		Jamb-up Gaskets		Foam strips
	S-shape	V-shape	Flexible	Rigid	
Cost	$0.17–0.65/ft	$0.20–0.70/ft	$0.14–0.42/ft	$0.30–0.85/ft	$0.05–0.70/ft
Type					
Compression	yes	yes	yes	yes	yes
Sliding	yes	yes	not as good	not as good	no
Application					
Window	yes	yes	yes	yes	yes
Door	yes	yes	fair	fair	yes
Threshold	no	no	no	no	no
Effectiveness	good	good	good	good	fair to good; better for doors
Durability	good; adhesive-back, fair	good; adhesive-back, fair	fair	very good	fair

SOURCE: Madison Gas & Electric

While window quilts can be very effective insulators and draft-stoppers, they have several drawbacks. First, their effectiveness depends entirely on how diligent you are in using them. Second, they eliminate the view through the window and block natural light. Third, good-quality quilts are expensive. In fact, some of the lines I've seen advertised are almost as expensive as new windows!

In warmer climates, window treatments aren't selected so much for their insulating or air-sealing value as they are for their ability to reflect heat away from the glass, which cuts air conditioning costs, and the ability to filter out ultraviolet light, which cuts down on fading in fabrics.

In a world of changing seasons and unpredictable weather, window shades and quilts need to be as versatile as possible. Some styles have both insulating and sun-screening properties, and come in a beautiful variety of colors. Shades are also valuable in ensuring privacy when you want it and helping to screen out unwanted noise.

As long as your primary windows are still serviceable — that is, structurally sound and fitting snugly in their frames — I

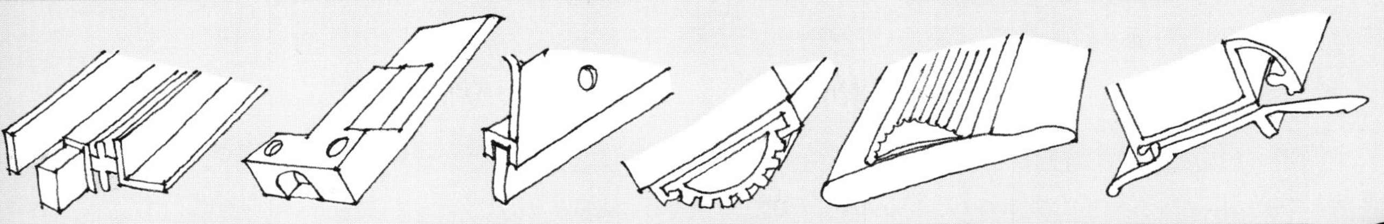

Magnetic Vinyl	Pulley Seals	Door Sweeps	Door Shoes	Vinyl Bulb Threshold	Interlocking Threshold
$1.05–1.50/ft	$1.50/pair	$3.00–7.00/door	$4.00–9.00/door	$5.00–15.00/door	$5.00–15.00/door
yes	N/A	yes	yes	yes	yes
yes	N/A	no	no	no	no
not double hung	yes	no	no	no	no
yes	no	no	no	no	no
no	no	yes	yes	yes	yes
very good	good	fair	good	good	excellent
very good	very good	good	very good	good	excellent

wouldn't recommend buying new ones for the sake of energy savings alone. It's when old windows get really ugly — beyond the saving power of a scraper and paint brush — and begin to cause comfort problems that the equation usually tips toward buying new windows.

If you do elect to buy new windows, be sure to cash in on recent advances in glazings and frame technology. For most remodeling jobs, I'd recommend double-paned windows with Low-E (low-emissivity) glass and an argon gas fill (instead of air) between the two panes. This combination achieves an insulating value of about R-4, compared to R-1 for a conventional single-pane unit.

Low-E, gas-filled windows may cost more up front, especially if you're ordering custom-built shapes, but they'll repay you many times over in comfort and economy. The gas buffer and heat reflective coating keep the surface of the glass warmer in the winter and help to reduce heating costs. You'll also experience less condensation on the glass and fewer uncomfortable drafts around the window.

Another good choice, though more expensive, is Heat

Mirror glass, which combines R-4 insulation with solar heat–blocking capabilities.

Windows with even higher insulating values are now appearing on the market, thanks in part to the development of new types of high-performance glass. Superglass, which employs a high-tech film manufactured by Southwall Technologies, is rated at R-9. When solar gain is taken into account, the glass can actually rival the performance of a R-17 wall! (See chapter 5, page 65.)

Keep in mind, however, that glass is only one of many components in a window. The quality and fit of the frame, sashes, weatherstripping, and hardware are equally important. So be sure to inspect the window before you buy it. Does everything fit? Wood windows should demonstrate fine workmanship, with the corners in plane and sturdy hardware throughout. Look for high-quality weatherstripping that runs in a continuous band around the window — this provides a better seal than segmented strips.

Good vinyl windows are comparable to wood in insulating power and require a lot less maintenance. But the quality of the vinyl extrusions — that is, the plastic members used to build the window — can vary a lot from one manufacturer to the next. When comparing vinyl windows, ask to see some sample cross-sections of the extrusions used in the window. A good-quality extrusion will have thick, rigid walls and a number of closed inner chambers or cells for added strength. Generally speaking, the thicker the walls and the more inner cells the extrusion has, the better.

I think it's still advisable to steer clear of darker colors, which absorb solar heat and aggravate vinyl's tendency to expand and contract. If you're shopping for extra-wide windows, bays, or bows, look for aluminum or steel reinforcing inside the extrusions. And remember: No matter how alluring the price may seem, *cheap* vinyl windows are seldom a bargain.

A group called the National Fenestration Rating Council (NFRC) is developing a performance rating system for new windows — similar to the miles-per-gallon labels you find on new cars — that will help shoppers wade through the confusion of numbers and performance claims that currently muddy the market.

Until the NFRC's consumer labeling program is ready, I think one of the best overall measures of a window's craftsmanship is its air-infiltration rate. The industry standard is expressed in CFM (cubic feet per minute) per linear foot of crack; the lower the number the better. A well-built casement or awning window might have an air-infiltration rating of .10 or less. The numbers tend to be slightly higher for double-hung and sliding units, because they're generically leakier.

If you live in the northern part of the United States, where heat loss through windows is the overriding concern, shop for a window (not just the glass, but the overall unit) that has a low "U-value" rating. (U-value is the inverse of the previously defined R-value, that is, 1 divided by the R-value.) U-values, which range from about .70 for the worst-performing windows to around .22 for the best, are an indicator of how much heat will be lost through the window.

People who live in the southern part of the country, where the primary goal is to keep solar heat from entering the house, should pay close attention to the window's "shading coefficient." (The shading coefficient is an indication of the window's ability to block solar heat, with the good-to-poor scale running from about .45 to 1.0.)

If your shopping list also includes new doors, opt for a style that's got plenty of insulating power, good weatherstripping, and a sturdy, tight-fitting latch. If there's glass in the door, make sure it's double-paned and tightly sealed. A good storm door serves to reinforce the insulating and air-sealing features of the primary door, and also protects its finish.

Some of our favorite "This Old House" projects have included the addition of air-lock entryways or, as they're more commonly known in New England, "mudrooms." These usually small vestibules provide an important buffer between the main part of the house and the outdoors, ensuring that cold blasts of winter air or summer heat can't sweep directly into your home. Besides saving you money on your utility bills, mudrooms provide a handy transitional area where the family can shed its muddy boots and wet raincoats.

Since it's usually impossible to equip every exterior door with its own mudroom, our strategy has been to locate the mudroom behind the door that gets the heaviest traffic.

Depending on how your space is laid out, you may be able to design and equip a mudroom so that it serves several useful purposes. In a recent project in Lexington, Massachusetts, we converted one side of a two-car garage into a multifaceted space that served both as mudroom and pantry, and also housed the stairway leading to the floor above.

Once you've completed a systematic weatherization of your house, you'll find that it's a lot more comfortable place to live in. And you'll enjoy savings up to 35 percent on your heating and cooling bills.

It's possible — even for a do-it-yourselfer — to tighten a house to the point where indoor air pollution and moisture problems

Shopper's Guide: Insulation

Insulating Material	Form	R-Value per inch	Thickness (in Inches) Needed for: R-11	R-19	R-30
Fiberglass	Loose	2.5	4.5	7.5	12
	Batt	3.1	3.5	6	9.5
	High-density batt	3.5	3.5[1]	5.5[2]	8.5
	Board	4.0	3	5	7.5
Cellulose	Loose	3.7	3	5	8.5
Polystyrene					
Expanded (beadboard)	Board	3.5	3	5	9
Extruded (blue- or pink-board)	Board	5.0	2.5	4	6
Polyisocyanurate (Foil-faced)	Board	6–7	2	3	4.5
Phenolic Foam (Foil-faced)	Board	8	1.5	2.5	4
Inject-in-Place Foams[3]					
Air-Krete	Foam	3.3–3.9			
Tripolymer	Foam	4.6			
Polyurethane	Foam	7.5			

1. Yields R-15.
2. Yields R-21.
3. Injected into existing wall and roof cavities; professional contractors only.

SOURCES: *Journal of Light Construction*; Madison Gas & Electric Co.; Massachusetts Audubon Society; Owens-Corning Fiberglass.

occur. We'll talk more about this in the upcoming chapter on ventilation and air quality.

Insulation

As with airtightening, the best place to start insulation work is usually up in the attic. Chances are that you already have *some* insulation up there, but with the rising cost of fuel — especially electricity — and the nation's continuing dependence on imported oil, the U.S. Department of Energy keeps ratcheting up its recommended levels.

Using the map and chart on page 53, you can decide if you need additional insulation, and what kind of materials might suit you best.

Insulating the attic is a fairly straightforward job for a do-it-yourselfer, unless the house has a flat roof, mansard roof, cathedral ceiling, or an attic with permanent flooring. These structural features complicate the job and may call for a professional contractor, or at least professional consultation.

Fiberglass blankets are the most popular add-on insulation for do-it-yourselfers, because the rolls, which come in 16- and 24-inch widths, lay down nicely between the joists. If you find that the joists in your attic are irregularly spaced or that there are lots of obstructions between them, you can buy bags of loose-fill insulation instead and just pour it in. Again, fiberglass is the most common choice, but you can also buy loose-fill vermiculite, polystyrene, cellulose, and perlite. Sometimes it can be advantageous to combine blanket and loose-fill insulation on the same job.

Experience has taught me that attic work will go a lot faster and easier if you take the time to rig up some temporary flooring (planks laid across the joists) and lights. There's not much hazard in the job, really, unless you bang your head on a rafter (look out for nails jutting down through the roof!) or underestimate the itchy, lung-gagging, eye-burning capacity of fiberglass. In other words, be sure to wear a thick, long-sleeved shirt and gloves, protective eyewear, and a dust mask.

The trickiest part about attic insulation work is getting the plastic vapor barrier right. If you live in a cold climate like mine, where *heating* is the main concern, the barrier always goes down first, against the ceiling, so that it can keep the warm, moist air inside the house from escaping up into the insulation and roof members.

In a mild or "mixed" climate, where both heating and cooling are required for several months of the year, a vapor barrier may or may not be advisable. Ask a quality contractor or building inspector in your area about local codes and practices regarding the use and placement of vapor barriers.

For those of you who live in hot climates, where air conditioning is the overriding concern, it's probably best to omit the vapor barrier altogether.

Properly installed, a vapor barrier reduces energy costs and prevents moisture from condensing in the wrong places, where it can degrade the value of insulation, cause wood rot, or give you a bad case of bleeding walls and ceilings.

I think the easiest way to create a vapor barrier is to buy insulation that already has foil or paper backing attached. In our climate, here in Boston, you would place the backing down flush against the ceiling and staple it as snugly and evenly as possible along the joists on either side. Where one batt ends and another begins, tape the seam so that there's no break in the barrier.

Another way to create a vapor barrier is to cut strips of 4- or 6-mil plastic sheet and lay those between the joists, lapping a little extra up the side of each joist so there's a ready surface to staple. Unfaced batt or loose-fill insulation would then be added.

In all cases, the vapor barrier and insulation should extend out far enough to cover the top plate — that is, covering all of the

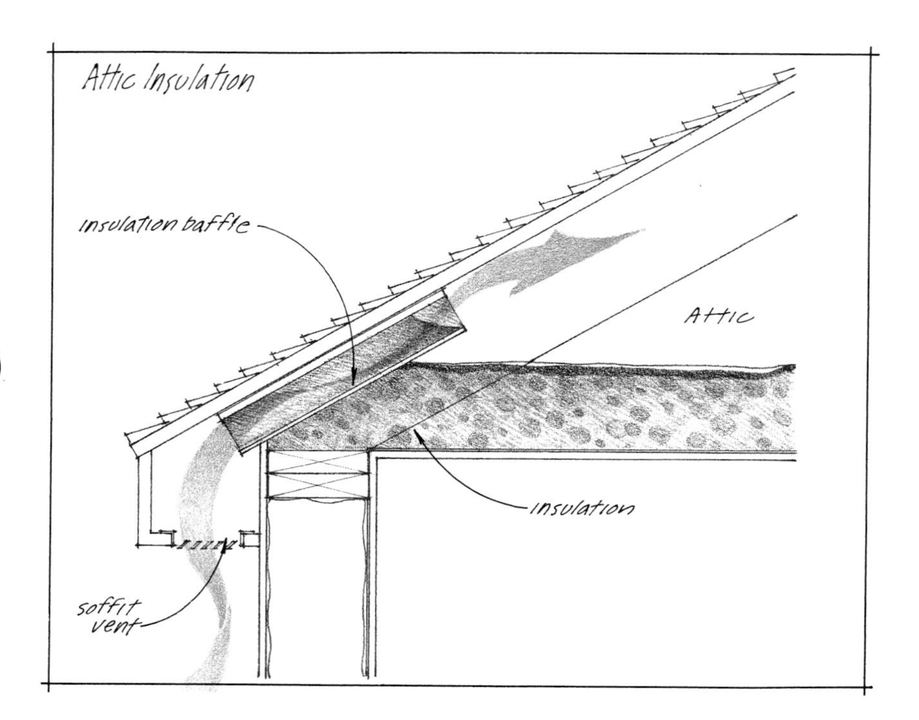

conditioned space below — but not so far as to block the soffit vents that run along the underside of the eaves. These openings help ventilate the attic and keep it dry. Store-bought or homemade baffles made out of fire-retardant cardboard or other inexpensive materials can be fixed around the soffit vents so that insulation can't choke the vent opening.

If you have recessed lighting fixtures jutting up into the attic space, check to see if they're code-marked "IC" (for Insulation Contact-rated), which means you can put insulation directly over and around the fixture without creating a fire hazard. If the light isn't marked that way, you'll need to frame in a little airtight box around the fixture so that insulation doesn't come in direct contact with it. The box can be built out of fiberglass ductboard and sealed with mastic. The height and width of the box should provide at least a 6-inch clearance to the light fixture.

When adding new fiberglass batts on top of old, you'll get best results if the new layer is laid perpendicular to (across) the one below. Be sure to use unfaced batts for this application so that you don't inadvertently create a second vapor barrier, trapping moisture in the insulation. If you're working with loose-fill insulation, you can spread it out with a board or garden rake, using a string line to get it level. Never pack or jam insulation into a cavity — compressing it only degrades its value.

●

Once the attic is done, consider insulating options in the main part of the house and basement.

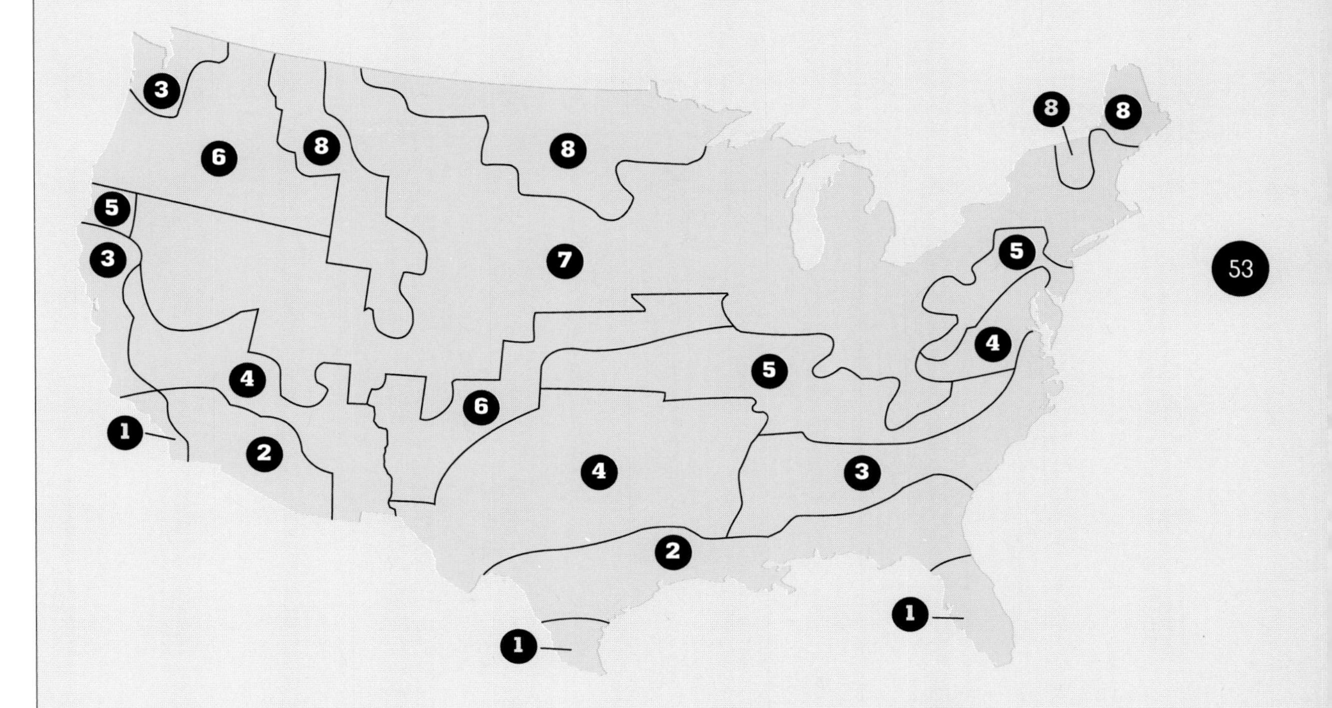

Recommended Total R-Values for Existing Houses for 8 Insulation Zones

Component	Floors over unheated crawl spaces, basements	Exterior walls (wood frame)	Crawl space walls	Ceilings below ventilated attics	
Insulation zone	All fuel types			Electric resistance	Fossil fuel
1	11	11	11	30	19
2	11	11	19	30	
3	19	11	19	38	30
4	19	11	19	38	30
5	19	11	19	38	
6	19	11	19	38	
7	19	11	19	49	38
8	19	11	19	49	

1. These recommendations are based on the assumption that no structural modifications are needed to accommodate the added insulation.
2. R-Value of full wall insulation, which is 3 1/2 inches thick, will depend on material used. Range is R-11 to R-13. For new construction R-19 is recommended for exterior walls. Jamming an R-19 batt in a 3 1/2-inch cavity will not yield R-19.
3. Insulate crawl space walls only if the crawl space is dry all year, the floor above is not insulated, and all ventilation to the crawl space is blocked. A vapor barrier (e.g. 4- or 6-mil polyethylene film) should be installed on the ground to reduce moisture migration into the crawl space.

NOTE: For more information see: DOE Insulation Fact Sheet (DOE/CE-1080), U. S. Department of Energy, Technical Information Center, P.O. Box 62, Oak Ridge, TN 37830.

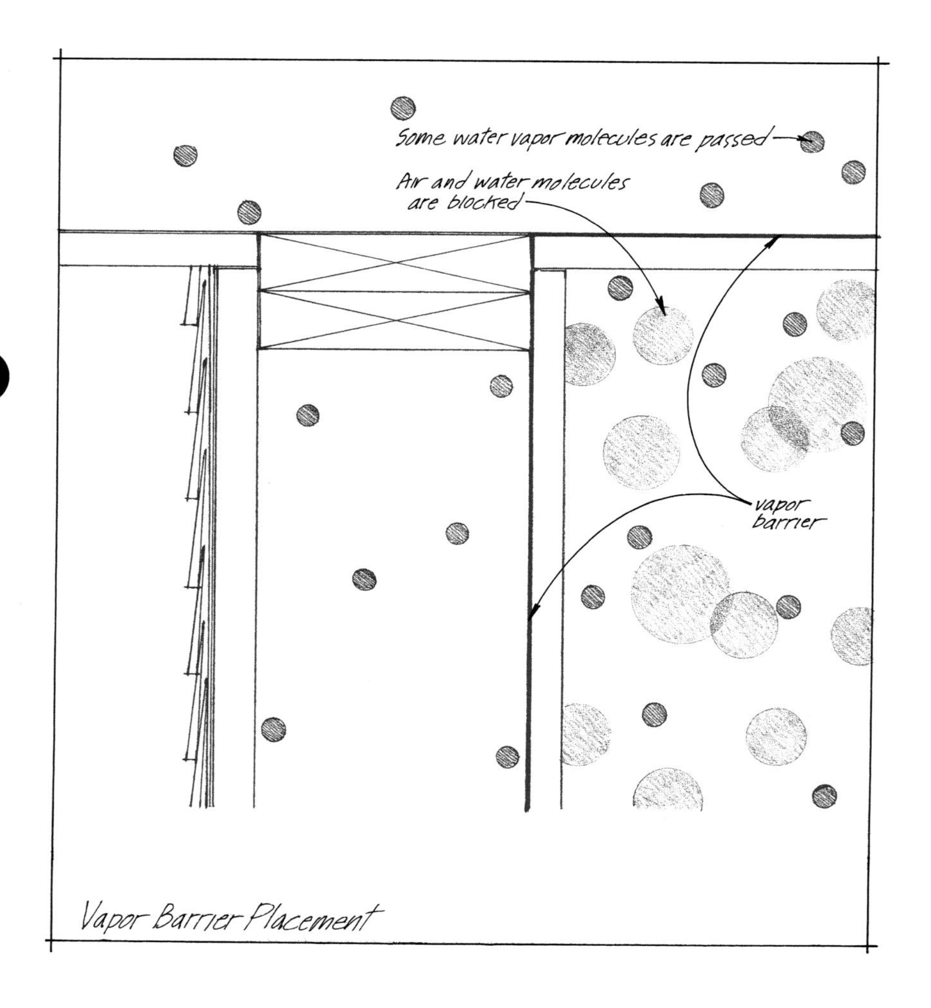

Vapor Barrier Placement

One way to add insulation to existing walls is to have a professional contractor pump it full of cellulose, fiberglass, or foam. This can be an especially smart move if you live in an older home with little or no wall insulation.

Some of you may remember a "This Old House" episode — circa 1990 — when we showed how cellulose insulation can be blown into existing walls without much muss or fuss. The insulation values in the walls of that house — an old triple-decker located in Jamaica Plain, Massachusetts — went from R-2 to R-14 almost overnight. It's the kind of smart energy retrofit that pays off in a lot of different ways.

First of all, those warm, insulated walls and relatively draft-free rooms made life immeasurably more comfortable for Hazel Briceno, the owner, and the tenants who occupied the other two floors of her house.

Second, Hazel will recover the $2,500 she invested in the job many times over. That's because the added insulation enabled us to install a smaller, less-expensive boiler than would otherwise have been possible and because she'll be enjoying lower fuel bills for years to come.

Finally, there's an environmental payoff, not just for the folks in Jamaica Plain, but for the nation and world around. Just think of all the fuel that *won't* have to be produced, transported, and combusted thanks to Hazel's insulation, and of all the contaminants that *won't* have to go up the flue. Less, in this case, really does turn out to be more.

Insulation contractors figure the price of the job by the square footage of the stud bays to be filled. An average house might run $1,000–$1,500 for blown-in fiberglass. On the Briceno project, the insulation was blown in from the inside of the house because we were planning on doing extensive re-plastering anyway so there was very little added cost in drilling holes in the inside walls. It's more common to drill the holes and blow in the insulation from the outside, and then plug the holes in the siding.

There are various kinds of blown-in-place and foam-in-place insulations to choose from, including cellulose, fiberglass, rockwool, polyurethane foam, Tripolymer foam, and Air-Krete, which is made from a special type of cement. All of these depend heavily on the skill and integrity of an experienced contractor.

Maybe I'm old-fashioned, but I think blown-in cellulose or fiberglass are still the best choice for houses. They don't have the higher R-values that some of the foams offer, but neither do they cost as much. And foams are more exacting in their application, requiring just the right pressure, temperature, and chemistry.

There's been some concern among environmentalists that Tripolymer foam could release formaldehyde gas into the home over time, though the reports I've seen were long on worried speculation and short on documentation. Another type of insulating foam — urea formaldehyde — was outlawed because of similar health concerns.

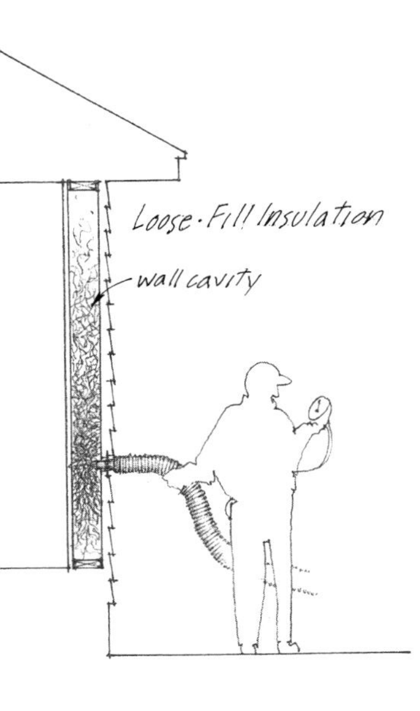

There are also environmental concerns about polyurethane foam, which gradually outgases chlorofluorocarbons (CFCs), a prime suspect in the depletion of the earth's ozone layer. (By the way, these same CFC worries apply to insulating foam board — including polystyrene, polyurethane, and isocyanurate — putting intense pressure on manufacturers to find alternative materials and manufacturing methods.)

Air-Krete, a cementitious foam that won't burn, has established itself as an environmentally sound product. But it has to be carefully formulated at the job site and is tricky to apply, especially in closed wall cavities.

Because it's so difficult to blow or foam insulation into a wall cavity without leaving voids, quality contractors sometimes use an infrared camera to scan the finished job and reveal the voids. Since infrared scanners cost about $30,000 a pop, a contractor has to be pretty serious about his work to own one.

Another way to add insulation to existing walls, as we've demonstrated on the show a couple of times, is to frame in a new stud wall or mount furring strips on the inner side of the wall so that a new layer of rigid or batt insulation can be added. To my way of thinking, this type of insulation retrofit usually isn't cost-effective unless it's done in tandem with other interior renovation work. And of course, you have to sacrifice some floor area, since you're building in from the wall. But the results can be very satisfying.

For example, if you're going to remodel the basement, take advantage of the opportunity to add new insulation (and vapor barrier if needed) before the drywall and paneling go up. If you're doing outside excavation work around the basement wall — to unplug a footing drain, for example, or repair the foundation — consider adding new insulation to that outside wall while you have the chance. (Be sure, in every case, to deal with leaks and high humidity in the basement *before* you add any insulation.)

Likewise, if you're going to put up new siding, why not tighten up your house and/or add insulation at the same time? Once you've got the old siding stripped away, it's relatively easy and inexpensive to install an airtightening or insulating wrap before the new siding goes on. (In some cases, the new siding itself may have considerable R-value.)

Some of these "house wraps" and exterior insulating products are designed to stop air movement through the wall, but not moisture. Others stop both. It's critical in selecting the right product that your supplier or contractor understands the climate and the characteristics of the existing wall, and most important, how the new "wrap" is going to affect the movement of moisture through the wall. If the wrap ends up trapping unwanted moisture in the wall cavity, the results can be disastrous. I know of one house in Maine that was so full of rot — at the tender age of four — that it finally had to be bulldozed.

The illustration on page 54 details some other areas in the home that should be insulated, including sill boxes, floors over unheated spaces, and ductwork and pipes that run through unheated areas.

I'll close here with a short reminder about the Law of Diminishing Returns. While it always pays to do a good, thorough job with airtightening and insulation work, the Law of Diminishing Returns tells us that as you invest more time, money, and materials into a

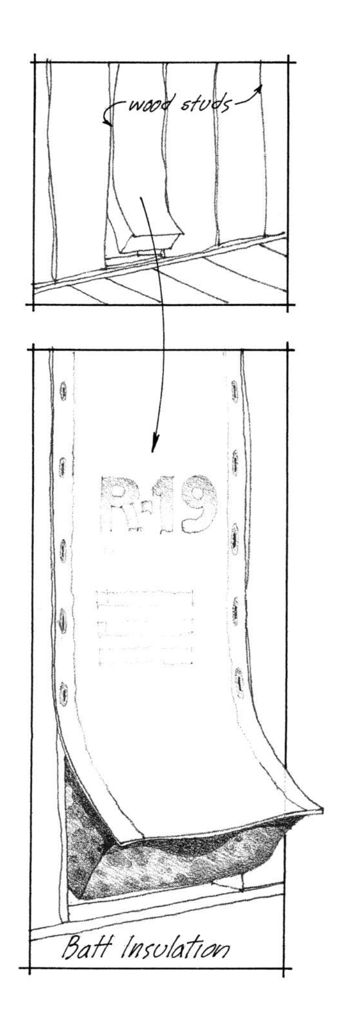

Batt Insulation

High-Tech Helpers

Though this may look like a radioactive roof photographed somewhere downwind of Chernobyl, it's actually a fairly typical house viewed through the eye of an infrared scanner. These sensitive devices can detect and measure very small temperature variations on the surface of a wall or roof. Infrared scans are usually done at night or on a very cloudy day and work best when there is a 25-30°F difference between the temperature inside the house and out.

Scans can reveal spots where heat is leaking through the home's envelope, including insulation voids down to the size of a baseball. These faults show up as bright spots on the scan. The warmer the spot, the brighter its color. The house pictured here, for example, is losing a lot of heat through the roof, because there's little or no attic insulation.

The practical value of infrared scans was shown when a nonprofit group called Rhode Islanders Saving Energy tested 504 houses that had had loose-fill insulation blown into their walls. The infrared tests uncovered large voids in the insulation on many of them — in fact, about half the houses flunked the test.

One contractor, who had insulated a house with 2,392 square feet of wall area, was reluctant to return after the infrared scan revealed 234 square feet of voids. He told the homeowner: "Well, 90 percent of the job is pretty good, isn't it?"

The homeowner responded by saying that he was only going to pay 90 percent of the bill. That brought the contractor back on the job pronto.

This photograph, taken with an infrared camera, reveals hot spots (red) where heat is leaking through the house's envelope.

If your insulation contractor doesn't have an infrared scanner — most won't, since it's a $30,000 item — you may want to hire an independent scan, which costs from $75 to $250. By banding together with a neighbor or two who also want a scan, you can usually get the inspector to give you a group discount on the price.

If you tell the insulation contractor beforehand that you're going to have an infrared scan done, and that he'll be expected to come back and fill any significant voids, and that those terms are going to be written into the contract, he might do the job right the first time. (Look over the contract carefully to make sure the fine print doesn't exempt the contractor from callbacks resulting from a scan.)

To find a themographer in your area, check under "Infrared Inspection Services" in the yellow pages or contact The Infraspection Institute, 33 Juniper Ridge, Shelburne, Vermont 05482.

particular aspect of the job — beyond a certain, judicious point — the value of the payback starts to taper off.

For example, if you already have insulation in the attic with an R-value of 22 — but none in the basement — you'd get a better return on your work by adding insulation to the basement rather than piling more into the attic. In other words, you've already saved the lion's share of what's possible to save up in the attic, while the potential energy savings and comfort improvements in the basement haven't been tapped at all.

The best way to make decisions about diminishing returns is to think about your house as an integrated system, with many different parts that need to be improved in an evenhanded manner. Common sense can go a long way in telling you how to allocate time and resources so that one part of your home's envelope doesn't end up perfected and another neglected.

●

Using the Site and the Sun

Not many folks have the chance to build a house from the ground up. If the opportunity comes along at all — to shape dreams into wood and brick and glass — it's usually once in a lifetime.

But even if you don't plan on ever building a new house, you can use the ideas and principles described in this chapter to build smart additions and generally improve the quality on other types of renovation work.

Back in 1984, the "This Old House" crew set out on a bold new course. Instead of tackling another renovation project, we decided to try our hand at new construction. *The All-New This Old House,* as it was called, was actually commissioned by the Boston Edison company to showcase state-of-the-art design, materials, and building techniques, and to test some promising technologies of the future. With those goals in mind, Boston Edison dubbed their project the "Impact 2000 House."

We constructed the three-bedroom, 2,800-square-foot house on the side of a hill in Brookline, Massachusetts. As you can see from the photograph, the south-facing wall and roof are heavily glazed and oriented to collect lots of wintertime light and heat. The northern face of the house, which has relatively few windows, is nestled back into the hillside where it's shielded from winter winds and benefits from the year-round insulating power of the earth.

When you think about it, orienting a new house or addition so that it can operate in harmony with the sun, the wind, and the earth is nothing but good common sense. The Pueblo Indians and their ancestors, the Anasazi, knew this when they positioned their dwellings in the hollows of south-facing cliffs. So did the farmers and ranchers who opened up the American prairie, planting trees and hedgerows around their houses to break the winter wind and provide islands of cooling shade in the summer.

As demonstrated in some of the nineteenth-century and early-twentieth-century houses we've worked on, early-day architects and builders knew how to lay out a floor plan so that windows, doors, transoms, and stairwells could intercept and channel prevailing breezes to ventilate and cool the house naturally. They also understood — perhaps through intuition more than science — the

In 1984, "This Old House" collaborated with Boston Edison to build the "Impact 2000 House" in Brookline, Massachusetts. By combining passive solar design with state-of-the-art HVAC systems, we produced a home that is extremely energy efficient, quiet, easy to clean, and — according to its owners — a sheer delight to live in.

value of thick masonry or adobe walls and massive central chimneys as architectural features that could moderate indoor temperature swings. Sometimes they'd even include a glassed-in porch or conservatory on the south side of the house, which provided the owners with lots of natural light, a measure of solar heat, and the chance to have vine-ripened tomatoes in February.

But cheap oil, natural gas, and electricity sidetracked us for a while, encouraging a lot of architects, builders, and homeowners — not to mention politicians — to make shortsighted choices. Hundreds of thousands of houses were built during the 1940s, '50s, and '60s with no regard at all to how the sun might cross their roofs and windows. Building lots were stripped bare of trees and other vegetation with nary a thought about windbreaks or shading, let alone the environmental impact. And who had time to ponder the influence that seasonally changing winds might have on the owner's comfort and utility bills?

To compensate for our foolishness in the siting and construction process during those lost decades, we slapped in mile after mile of baseboard electric heating and routinely oversized our furnaces, boilers, heat pumps, and air conditioners by a factor of 50, or even

100 percent. (Many of these systems, by the way, are still out there clunking along.) It wasn't until the price of fuels began to climb — and finally soared — that our national folly became apparent.

Which brings me back to the house in Brookline. In keeping with the wisdom of the ancients, it was first and foremost designed to work in harmony with the geography, vegetation, and climate at the site.

As a second principle of design, its foundation, walls, and roof were tightly built and superinsulated — R-68 in the ceiling; R-36 and R-38 in the walls and floors, respectively — so that the home's heating and air conditioning requirements were minimized from the start. The home's high-performance windows, forerunners of the one shown on page 65, were chosen to trap solar heat and provide R-4 insulation, about twice the value of ordinary double-pane units.

In my opinion, these twin principles — proper siting and a tight, well-insulated envelope — are the necessary starting points for a truly comfortable and affordable house. And building a house or addition in this fashion won't add much, if anything, to your net cost. Proper siting doesn't demand anything but a little site exploration, some knowledge of the climate, and a few hours of intelligent thought. If your architect or builder can't handle it, you'd be well advised to shop around.

The extra materials and labor needed to build a tight, well-insulated envelope are immediately offset by the savings you get in downsizing the heating and cooling gear, and vastly compensated over the longer term with lower utility bills.

Perry Bigelow, a Chicago-area builder who specializes in energy-efficient construction, has reduced the heating load in his new houses to the point where they can be heated by an oversized water heater. This is accomplished by installing a separate piping loop between the water heater and the fan-coil unit. When the thermostat calls for heat, hot water is pumped out of the water heater and through the coils. Air blown across the coils is heated and distributed through the home's ductwork.

Perry's approach costs $1,000 to $1,500 less than buying a traditional heating system, which pays for much of the envelope work that cut the load in the first place. Perry guarantees buyers that their annual heating bill won't be more than $200 or he'll step up and pay the difference. So far — and he's been in the business for a good many years now — he hasn't paid a dime.

Some builders take energy-efficient construction techniques even further, to the point where the house is essentially heated by its internal gains, that is, the heat generated from electric lights, appliances, and human activity.

The counterpoint to all this is that if you don't properly site the house when you build it, or build a weak envelope, it's going to cost you an arm and a leg to correct the situation later on — if in fact it's correctable at all.

Letting the Sun Shine In

With its 440 square feet of south-facing glass and large amounts of thermal mass (select portions of the interior and exterior walls are built of concrete block), the Brookline house is a passive solar house in the truest sense. It was designed and built to capture as much of the sun's free heat as possible in the winter and to benefit from natural cooling in the summer.

The wintertime sun, which hangs low in the south at our latitude, can pour directly through the south-side windows of the home. Inside, the thermal mass works like a battery, storing up solar heat until it's needed, then gradually releasing it, which reduces day-to-night temperature swings.

During the summer, when the sun is high overhead, the home's south-facing windows are shaded by permanent overhangs. Some of the windows are also equipped with operable sun shades. Both the overhangs and the shades are important features during the cooling season, so that you don't have incoming solar

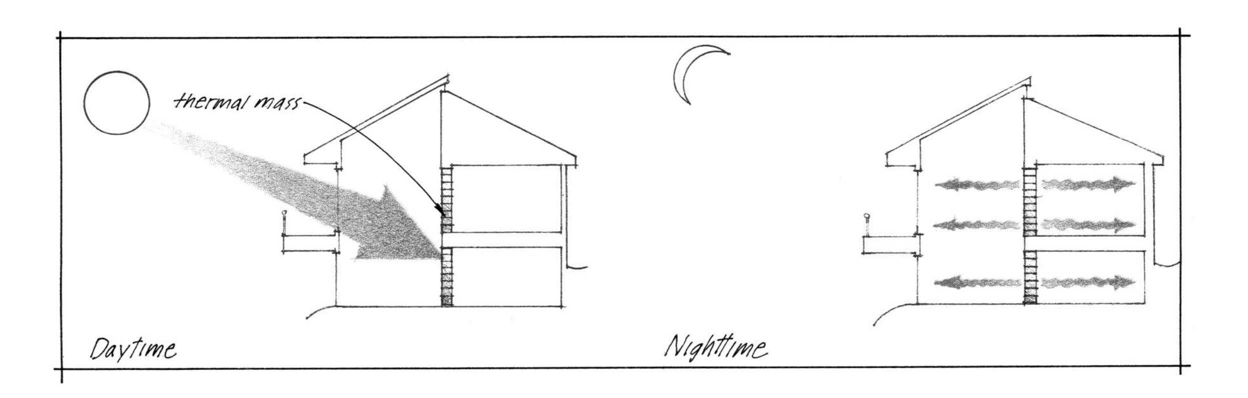

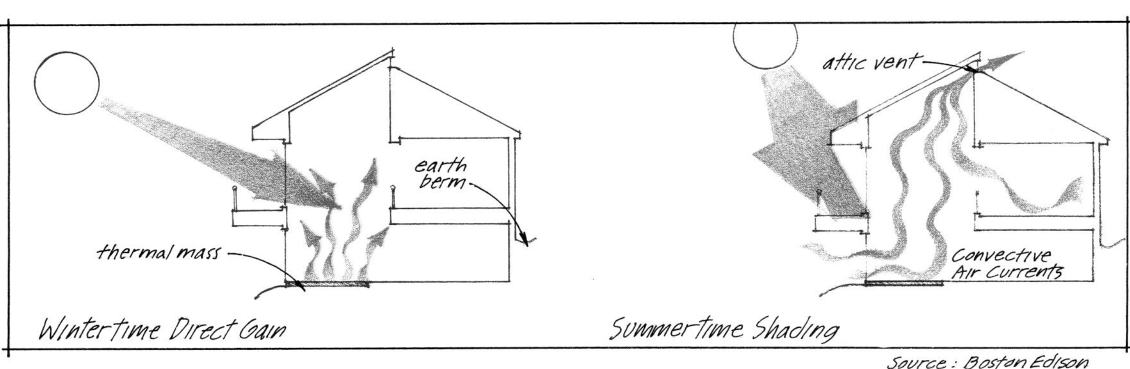

Source: Boston Edison

heat putting an extra burden on the air conditioning system. The thermal mass, reversing its heat storage role, now helps the house stay cool.

Though it sounds pretty novel, none of this is really new. In fact, Socrates, who lived some 2,400 years ago, is generally credited with being the first passive solar architect. He achieved what he called the "comfort principle" — that is, staying warm in winter and cool in summer — by designing houses with large south-facing windows, equipped with balconies or awnings, and sheltered on the north.

Of course, you don't have to build a full-blown passive solar home to take advantage of the sun. Builder Doug George, whose work is shown on page 77, meets 20 to 30 percent of the heating load in his houses (and sometimes more) with solar energy, using only normal amounts of south-side glass and no extra thermal mass.

"Because we build our houses so tightly and superinsulate them, the solar gain we do get meets a very large percentage of the heating load," Doug says. "Superinsulation and passive solar make a beautiful marriage, even when the design isn't consciously or conspicuously solar."

As a rule of thumb, the proper amount of south-side glazing is calculated as a percentage of the home's total floor area. If the area of south-facing glass exceeds 7 percent of the total floor area, the design should begin to include extra amounts of brick, tile, stone, or other types of thermal mass so that the house won't overheat. Other key features in a good solar design might include overhangs, sun screens, insulating blinds, existing or newly planted vegetation, and a combination of natural and mechanical ventilation.

While a "sun-tempered" house like the kind Doug George builds might meet 20 or 30 percent of its heating load with solar gain, a fully developed passive design could easily achieve 50 percent or more. The house in Brookline, for example, which has south-side glazing equal to 16 percent of its floor area and plenty of extra mass in its walls and floors, meets more than 60 percent of its heating load with solar without increasing the summertime air conditioning load.

Mechanical Systems

After reducing the heating and cooling loads as much as possible, through the techniques described above, architect Steven Strong, head of Solar Design Associates, and the planners at Boston Edison turned their attention to the remaining heating and air conditioning load. They decided to use a geothermal, or ground-coupled, heat pump.

Unlike air-to-air heat pumps, which draw their heat from the air, geothermal designs draw their heat from the earth itself.

At the Brookline house, which has an open-loop geothermal system, water is pumped up from the bottom of a 750-foot water well, routed through the heat pump, and then discharged back into the top of the well. In other words, the well serves as the system's year-round heat source and heat sink.

Since the temperature of soil and water below the frost line remains stable at about 50°F in our climate, ground-coupled heat pumps can outperform the air-to-air variety by a healthy margin.

When the air gets too cold for air-to-air heat pumps to run efficiently, and they revert to being electric resistance heaters, ground-coupled units keep right on humming.

In the summertime, the ground-coupled heat pump reverses its function, extracting heat out of the house and dumping it into the water well, providing the Brookline house with about four tons of air conditioning.

Another way to configure ground-coupled systems — which I think is preferable — is to use a closed loop of pipe buried in the ground. Instead of cycling lake or well water through the system, a mixture of water and antifreeze is pumped continuously through the closed loop. During the winter, the heat pump extracts heat from the ground and pumps it into the house. In the summer, the process is reversed.

Among the Brookline house's many other notable features is its solar roof, which provides both solar-heated water (85 percent of the load) and solar-generated electricity (30–35 percent of the load). The house is also equipped with a heat-recovery ventilating system, two heat-circulating fireplaces with dedicated air supply vents, and energy-efficient appliances throughout.

In 1986, when Boston Edison ran its last public tour through the house, it was sold to Justin and Genevieve Wyner, who have lived there ever since.

"Not only is this house very, very efficient from an energy standpoint," says Genevieve, "It's also much more comfortable, quieter, and easier to clean than any of the other houses we've lived in. The design manages to be extraordinarily beautiful even as it takes full advantage of what nature has to offer."

Justin, who has worked closely with Boston Edison in monitoring the performance of the passive solar and mechanical HVAC systems, says that the house is "substantially" less expensive to operate and easier to maintain than the several other houses he's lived in.

What the Site Suggests

The design for a new house or addition should in every case evolve from the climate and geography at the site, as well as the budget and personal tastes of the owner. The examples I've selected here — one each from New Hampshire, Florida, and Colorado — show how three prize-winning builders responded to the particular challenges they encountered at the site. Other designers might have responded differently to the same set of circumstances with equally smart and attractive results.

No matter where you live and what kind of architectural style you're after, the process of design and construction is always one of compromise. There is no single, guiding principle that must in every instance be applied — except for conscientiousness in planning and execution.

Here in New England, for example, it might seem the height of foolishness to build a house with broad expanses of north-facing glass — until you learn that the view to the north overlooks a spectacular stretch of the Maine coast (while to the south the view looks down on a seedy gas station on Highway 1!). So what's a good designer to do? Open up that view to the north, of course, but *compensate* — that's the magic word — with high-performance glass, insulating shutters, air-lock entryways, and/or other practical measures. You might call it watering down the foolishness.

In any case, it's the architect's and builder's responsibility to consider the constraints and priorities at the site and come up with a good — not perfect, mind you, but good — solution.

That's the basis on which the three case studies, starting on page 77, were chosen. They all represent good design solutions put into practice by good builders.

Remodeling

Because of the nature of the work, it's usually a very difficult and expensive matter to retrofit (that is, retroactively fit) a house with superinsulation or passive solar. As we discussed in the preceeding chapter, there *are* cost-effective ways to tighten up and re-insulate a house. But as a practical matter, the retrofit will never rival the airtightness and insulating qualities — nor the comfort and economy — of the new houses we're showing here. The limitations for adding passive solar to an existing home are even more pronounced.

The three most popular ways of adding a solar dimension to an existing home are by adding new windows, skylights, or a sunspace addition.

So long as the total glass area on the south side doesn't

Anatomy of a Super Window

If you were to add up all the energy lost through windows in this country, it would just about equal the amount of energy flowing through the Alaska pipeline each year. But window frame and glass technology are getting better fast, as illustrated in this cutaway of a window unit made by Hurd Millwork Co.

The unit features the new Superglass System, created by Southwall Technologies, which combines two sheets of Heat Mirror–coated film inside two panes of glass. All three of the air spaces are filled with insulating, environmentally safe, argon gas. The edge of the unit is sealed with special low-conductive spacers.

The glass itself achieves a remarkable R-9, four times the insulating power of ordinary double-pane glass. When energy losses through the frame are considered, the window unit as a whole achieves R-4.6, twice the insulating performance of a standard double-pane window. Wintertime tests have shown that if you factor in the solar heat that the glass admits, which reduces the load on your central heating system, windows equipped with this type of glass can actually outperform an insulated wall.

Windows with Superglass cost 20 to 50 percent more than conventional windows, but deliver superior comfort and pay for themselves in energy savings in a few years. In addition to its comfort and energy advantages, the glass blocks up to 95 percent of outside noise and screens out (without any tinting) 99 percent of the ultraviolet light. This extends the life of curtains, carpets, and other sun-exposed fabrics and keeps them from fading.

High-performance glass opens up all kinds of new design possibilities for architects, builders, and homeowners, who have been limited in their use of glass because of its energy and comfort liabilities. In fact, until the advent of high-performance glass, some energy and building code officials were pushing to limit the amount of window area a house could have. With windows on the market that can outperform many walls, code writers will have to be sensitive not just to the raw area of windows, but to how those windows perform.

State-of-the-art builders are already mixing and matching different types of high-performance glass to achieve extraordinary energy and comfort results. For example, it might make sense to use a heat-rejecting glass on west windows, high R-value glass on the north- and east-facing windows, and clear double glazing on the south to let in lots of wintertime solar heat.

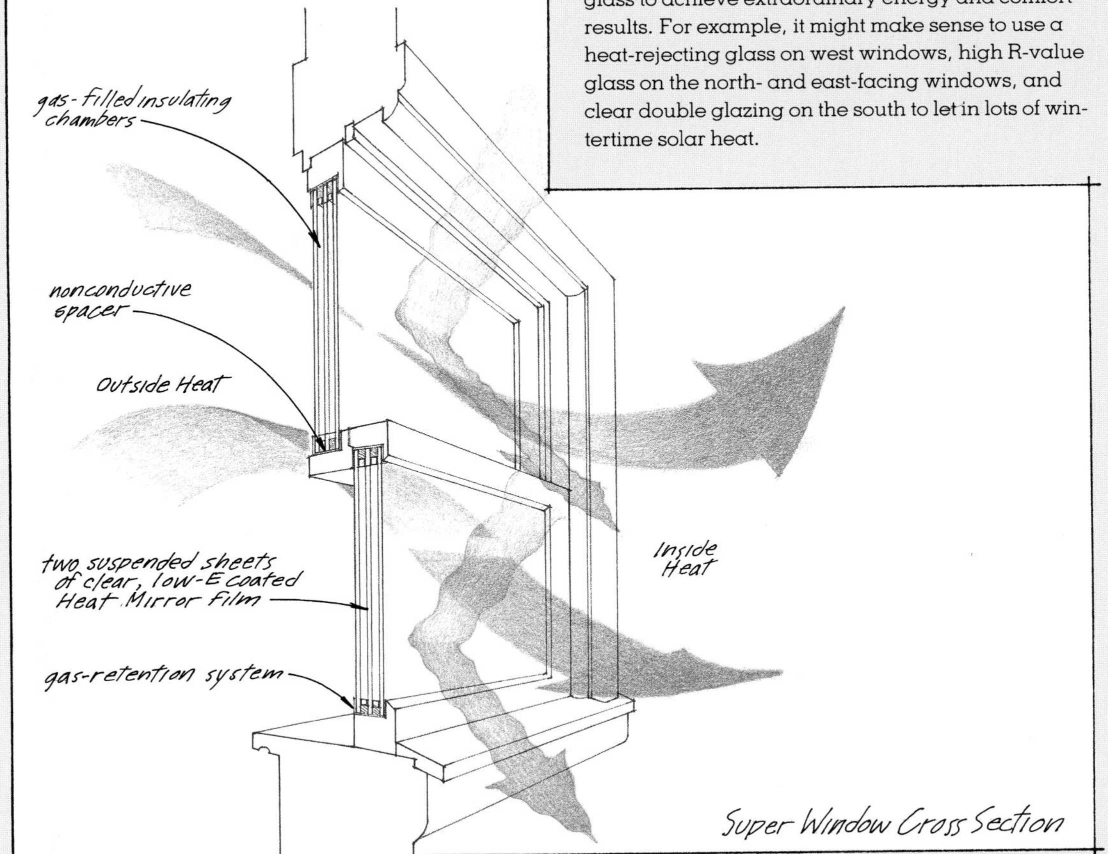

Super Window Cross Section

To keep the fierce Arizona sun away from the windows of our project house in Phoenix, we equipped them with heavy canvas awnings that add a nice aesthetic touch as well.

exceed 7 percent of your home's floor area, you shouldn't experience any overheating problems. Having said that, I hasten to add that some of the biggest summertime glare and heat problems actually come from west-facing windows (and to a lesser extent those on the east).

The key, very early in your planning, is to think about how new windows or skylights are going to be oriented to the sun's path through the sky and how that path is going to change with the seasons. Where will the sun be around December 21, when it traces its lowest arc across the sky? How about on June 21, the summer solstice? At what angle will light and heat strike the glass and enter the room at different times of the year and different hours of the day?

Adding glass to your home's envelope, whether it's new windows, skylights, or a sunspace, will always affect your heating and cooling loads — on the plus or minus side — and raise the amount of natural light inside your home. The obvious strategy is to maximize the solar gain during the winter, when the heat and light are welcome, and minimize it during the summer.

I've already shared some tips on buying new windows in chapter 4. But it bears repeating here that homeowners are no longer bound to one or two choices when it comes to glass. With so many different varieties of high-performance and tinted glass on the market, you can sit down with your builder or architect and choose one that's just right for the application. You might end up using two or three different varieties for a new addition or sunspace project, each one selected for its special properties.

One of the most effective ways to keep unwanted solar heat out of your home, and avoid the air conditioning penalty that comes

The flashing and caulking details around roof windows and skylights must be done with care if you want to keep cold air and water out of your home.

with it, is to block sunlight before it ever hits the glass. Overhangs, trellises, awnings, and exterior sunscreens can accomplish this nicely. Interior white-backed blinds, shades, and curtains are helpful too, but not as good as exterior control, because the sunlight has already penetrated the glass.

Of course, nobody does it like Mother Nature. If you can arrange to leave deciduous trees in the right place, or plant new ones, they'll leaf out in the summer to block the sun's rays, then dutifully shed their leaves in the winter so that sunlight can come beaming through. Remember, in your planning, that even a leaf-barren tree will block some light and heat to the glass, so be sure to use them sparingly.

Skylights

The first rule in choosing a new skylight is not to skimp on quality. Select, at a minimum, a good double-paned unit with Low-E glass. Make sure that the manufacturer guarantees the skylight against leaks, cracking, warping, and discoloration. Metal-framed skylights should have a "thermal break" built in so that they won't conduct a lot of heat through the roof.

In our project house in Arlington, Massachusetts, we built a small deck out from the master bedroom that's accessible through sliding glass doors. This lets sunlight and fresh air into the bedroom and provides the owners with an intimate balcony. The skylights mounted on the lower roof play into the sunspace below.

Second, the installation work should be first-rate — after all, it's your *roof* that is being opened up.

Last but not least, consider some type of high-performance glass. Heat Mirror insulating glass, which employs a specially coated film, produced by Southwall Technology, will admit light through the skylight even as it blocks most of the sun's heat. The glass itself has a clear appearance. Tinted glass is another option for skylights, provided you don't mind the permanent color.

If you opt for a skylight with ordinary double-pane glass, I'd recommend using an operable interior shade with it so that when the linoleum underneath the skylight starts to melt you can simply close the blinds. (If you want to go high-tech with the shades, you can get a motorized version with remote control. This is a nice touch if the skylight happens to be twenty feet above the floor.)

Putting in skylights will probably add a few extra bucks to your annual heating and cooling bills. While it's true that skylights admit solar heat in the winter (how much depends on the size and type of glass, and how it's positioned to the sun), those gains are usually offset by increased conductive heat losses through the glass and frame. After all, you've replaced a patch of insulated roof that was probably R-22 or better with a piece of glass that will range, de-

Prefabricated sunroom kits are available in an almost infinite variety of shapes and materials. This particular model, from Four Seasons, has an aluminum frame with the roof and side walls entirely of glass.

pending on the type, from R-2 to R-9. In the summertime, skylights, even when they're properly shaded, tend to put a little extra burden on your air conditioner.

But after all, people don't put in skylights because they're practical — they buy them because they're beautiful to look at and open up the dark interiors of our homes to natural light.

And as long as you get a good quality double-pane unit that's properly weather-sealed, and have an adequate way to shade it, the heating and cooling penalty won't be enough to worry about. On the other hand, if you low-ball the job and get a cheap, single-pane unit that's not properly sealed, you've suddenly created a leaky chimney right up through the middle of your roof. In that case, your local fuel oil dealer, gas company, or utility will probably send you expensive Christmas cards and nominate you "Customer of the Year."

Sunspaces

One of the more ambitious ways to add a touch of passive solar to your house is with a sunspace, which can be created by adding glass and mass to an existing room or can be built on as a new addition.

As with other passive solar techniques, this one's nothing new. In fact, George Washington had a sunspace on his home at Mount Vernon.

What *is* new about sunspaces is the wondrous array of space-age materials we have to build them with, including high-tech glazings, laminated wood beams, neoprene gaskets, long-lasting caulks and sealants that move with the structure as it expands and contracts, and aluminum extrusions that are thermally broken to improve their insulating quality.

Companies that manufacture off-the-shelf sunspace kits and components offer an enormous selection of styles, materials, and prices to choose from. In addition to the standard packages available, most manufacturers will customize a kit to suit your particular needs. Only in those instances where you have a truly odd-sized space or need special design features — such as historical details — would an architect be crucial to the job.

This custom-built sunroom was added to a Georgian colonial (facing page) in a style that fits well with the formality of the rest of the house. A gabled entry divides the sunspace into a potting/plant room and a sunny breakfast nook (right). The operable wood casement windows are clad with aluminum on the outside for durability and easy maintenance.

Since sunspace kits are modular in design — like Tinker Toys — you can choose just about any size or shape you want. In urban areas they're used to enclose intimate little balconies and sky-view terraces set high above the bustle. In suburban and rural areas they sometimes grow into enormous two-story affairs that house swimming pools and clusters of banana trees. Small, large, or in between, sunspaces can assume an angular or curvaceous look depending upon the configuration of walls, eaves, and roof you select.

When it comes to function, sunspaces can be most anything you like: a snazzy new family room; a horticultural wonderland, replete with African orchids; a handsome air-lock entryway into the rest of the house; a prolific solar heating machine; an in-house gymnasium with body-building machines and hot tub.

To get precisely what you want, you have to establish priorities early on in the project and stick to them. The *first* point of discussion with a sunspace dealer, builder, or architect should be *how you intend to use the space*. In other words, the function of the sunspace is what determines its form.

Most sunspaces are built — first and foremost — because they provide a comfortable and creative living space, adding elegance to a home and creating a relaxing ambience. While solar energy and horticultural concerns may play second and third fiddle in the design process, they shouldn't be disregarded entirely. What sunspace would be complete without a few plants to enrich the environment? And why not make the design energy-smart, so you can tap some of the sun's heat?

A well-conceived sunspace design can be multifaceted, creating the handsome living space you want, providing a healthy environment for many types of plants, and tapping the sun's heat to shave a few bucks off your utility bills.

As you shop around, you'll find three types of common structural components. Extruded aluminum — the best-seller — is strong, lightweight, and virtually maintenance-free. It comes in various anodized and baked-on enamel finishes.

Solid wood frames made out of redwood or cedar are usually more expensive than aluminum, but they're beautiful to behold and will weather well. Less durable woods, like pine and fir, need to be continually maintained to protect them from the elements. For homeowners who want to combine a handsome wood interior with a tough, low-maintenance exterior, wooden frames with aluminum or vinyl cladding on the outside are a nice choice.

Laminated wood beams — a recent innovation in the sunspace industry — are a third option. By gluing numerous thin sheets of wood together until they attain the dimensions of a beam, manufacturers can create long sturdy spans and graceful arches made out of mahogany, redwood, cedar, yellow pine, or fir. Laminated beams (and most other types of wood as well) need to be tightly sealed with marine varnish or other heavy-duty sealants to protect them from the sunspace environment, where high humidity, temperature variations, and ultraviolet light abound.

It is glass, of course, that makes the sunspace such a special environment to live in. Though double-pane Low-E glass is the industry standard, there are dozens of other glazing options around, including laminated glass, tinted glass, Heat Mirror, Superglass, and various acrylics. The type of glazing you select and its placement in the structure will depend upon the climate, orientation, and use of the space. For instance, solid end walls might be desirable on some sunspaces to provide privacy and/or better energy performance. In other cases all-glass end walls might be preferable.

Obviously, with such a wide range of sizes, shapes, and materials available, sunspace prices vary greatly. If a dealer or builder does the whole job for you from A to Z, expect to pay between $60 and $135 per square foot of finished floor space. At that rate, the price of a room-sized sunspace (10 × 12 feet) would range from $7,200 to $16,200. As a rule of thumb, laminated beam and wood-with-cladding sunspaces cost more than simple wood beam or extruded aluminum frames. Apart from the foundation and frame, the glazing, flooring, and window coverings you select are key elements in determining the final cost.

As with windows and skylights, the sunspace's orientation to the sun is your first key design decision. If you're seriously interested in tapping solar heat, due south is best. But the sun is pretty flexible — you can situate a sunspace 45° east or west of south and suffer only a 20 percent reduction in winter sunshine. In really hot climates, it might even make sense to locate the sunspace on the *north* side of the house.

In the late 1970s and '80s, when the modern sunspace industry really got rolling, one of the most common problems was overheating, even in cold New England.

I could always tell when one of my friends or customers had been saddled with a poorly designed sunspace — when I walked into it, there'd be a pair of dark sunglasses and a bottle of No. 20 sun-block lying on the table. The few surviving plants were of the hardy cactus variety, and even they looked wilted.

The problem with many of those early-day sunspaces was that they had too much glass (especially in the roof), too little thermal mass to store the heat, no way to ventilate the space properly, and no means, inside or out, of blocking the sun.

My advice, in a nutshell, is to beware of too much overhead glass, unless it's a high-performance type that can deal with a Fourth of July sun and/or is equipped with an operable sun screen. As a general rule, vertically oriented glass is the best year-round solution.

Second, make sure that there's enough internal mass in the room — be it floor tile, bricks, or water tubes — to moderate the temperature. Houses made out of brick or cement blocks have an especially effective option for thermal mass built right in; that massive side wall can become the partition between the sunspace and the house, absorbing solar heat in the winter and conducting it directly into the house.

Third, look for sunspace designs that have been pre-engineered to provide good ventilation. The best systems promote natural ventilation by admitting fresh air through operable windows and

As part of the Bigelow restoration, in Newton, Massachusetts, we converted an old screened porch into an all-weather sunroom. Notice the handsome cedar trim used to finish the interior. Since the sunspace was not designed to heat itself year-round, hydronic baseboard heating was provided.

vents (positioned low in the sunspace) and exhausting hot, humid air through elevated vents and operable skylights. Natural ventilation may need to be augmented with fans in some cases. No matter what the combination, a flexible ventilation system is *essential* for year-round comfort and economy.

Top-Flight Help

If you're interested in using passive solar or superinsulating building techniques for a new home or addition, I suggest that you or your builder contact one or more of the following groups. All are top-flight organizations, interested in promoting good construction practices.

Energy Efficient Building Association
Northcentral Technical College
1000 Campus Drive
Wausau, WI 54401-1899
(715) 675-6331

American Solar Energy Society
2400 Central Avenue G-1
Boulder, CO 80301
(303) 443-3130

Passive Solar Industries Council
1511 K Street, NW
Suite 600
Washington, DC 20005
(202) 371-0357

I'd also recommend leaving a wall between the sunspace and the rest of the house, so the space can be easily sealed off. In essence, the dividing wall works as a giant damper. On a sunny winter day, when the sunspace is collecting solar heat, the doors and windows in the partition wall are opened to admit free heat to the rest of the house. The heat moves inside via natural convection or by using a small circulating fan. If the sunspace chills down during a spell of cold, cloudy weather, and you don't want to heat it with your central system or a space heater, all you have to do is close it off.

A partition wall is particularly important if you're going to have a hot tub or a lot of plants in the space, because you don't want the humidity they generate infiltrating the rest of the house. For heating, cooling, and ventilating such a space, I'd recommend a separate zone, which can be operated independently of the rest of the house.

Even if you have a partition wall and good ventilation in your sunspace, you may want to add insulating and/or sun-blocking window treatments as an additional control element. Some homeowners like to live with their sunspace for a while — evaluating its performance through the different seasons — before choosing a window treatment system.

If you're going to incorporate the sunspace directly into your house — with no dividing wall — the insulating power and sun-blocking characteristics of the glass you choose become doubly important.

When we remodeled Doug and Sarah Briggs's ranch-style house in Woburn, Massachusetts, the new sunspace — built into the breezeway between the garage and house — was left open to the rest of the house. Because there was no way to isolate the sunspace, we used high-performance Pella window units that have operable sun shades *between* the two panes of glass. The sunspace was also equipped with copper-fin baseboard radiators so the space could be heated with the home's main boiler.

And, finally, my advice to adventurous do-it-yourselfers is — *don't!*

Installing a sunspace properly is detailed and challenging work, even for experienced do-it-yourselfers. You're talking about a full-blown addition constructed mostly out of glass, requiring foundation, framing, and glass-fitting work that's less tolerant to error than other types of building. Beyond that, there's probably electrical and plumbing work to consider. If you want to invest sweat equity, that's fine, but I'd recommend working alongside a professional. One way to do this is to have the dealer or builder put in the foundation and erect the structure, leaving the interior and/or exterior finishing work up to you.

These same warnings hold true for installing skylights and — to a much lesser extent — new windows. Make sure of your skills before you barge ahead.

●

A Tale of Three Builders: *The Seydler Home*

Who, What, and Where

- Owners: Craig and Joan Seydler
- Builder: Doug George Homes, Dover, New Hampshire
- Location: Rollinsford, New Hampshire
- Style: Superinsulated contemporary with three-car garage
- Size: 2,600 square feet of finished living space (plus 1,300 square feet of heated but unfinished basement)
- Price: $205,500 ($79 per square foot; excluding land and site development)

Construction

- Southeast orientation to provide wintertime solar gain
- 2x6 wall construction with insulated sheathing
- Continuous airtight gasket system
- Continuous seam-sealed exterior air-infiltration barrier
- Continuous seam-sealed interior 8-mil vapor barrier
- Full walk-in basement
- Insulated steel entry doors (R-12)
- Vinyl-clad windows with Low-E, argon-filled glass

Insulation

- Flat ceilings: R-70
- Cathedral ceiling: R-70
- Above-grade exterior walls: R-32
- Basement walls: R-23
- Under basement slab: R-5

Ventilation

- Operable windows throughout
- Whole-house mechanical ventilation with heat recovery
- Full soffit and ridge vent roof system

Heating and Hot Water

- Heating system: Direct vent oil-fired boiler (AFUE: 84%)
- Distribution: Copper-fin baseboard radiators
- Control: Two-zone Honeywell thermostats
- Annual heating costs (with fuel oil at $.80 a gallon): $280
- Hot water: Indirect-fired insulated tank (R-18)
- Annual hot water costs (with fuel oil at $.80 per gallon): $200

Special Features

- EPA-approved sub-slab radon gas mitigation system

- Direct-vent fireplace with airtight doors and damper
- 90 square feet of fireplace brick providing extra internal mass to store solar gains and modulate temperature swings
- Kitchen featuring down-draft cooktop with halogen burners

Builder's Comments

"By building a tight, superinsulated frame on the Seydler house, we are able to meet one-third of its heating needs with natural solar heat. These solar gains are achieved with normal amounts of quality windows positioned mostly on the south and southeasterly quadrants of the house.

"Another third of the heat is provided by occupant gains, that is, the heat generated inside the house by humans, lights, cooking activities, and appliances.

"The final third of the home's heating load is met by a high-efficiency, sealed-combustion boiler rated at 60,000 BTUs. It's noteworthy that a house this size, built ten years ago using conventional building practices, would have required a heating plant at least three times as large. With fuel oil priced at 80 cents a gallon, the Seydler's heating bill is only about $280 a year — not bad for a cold-weather climate like New Hampshire's.

"In considering overall value, it's important to remember that the same details that go into making a home extremely energy efficient also make the structure durable and long-lived."

— Doug George, builder

A Tale of Three Builders: *The Brame Home*

Who, What, and Where

- Owner: Richard Brame
- Architect: Rick Cowlishaw, Monument, Colorado
- Builder: Marvin Gobbels, Monument, Colorado
- Location: Monument, Colorado
- Style: Passive solar, superinsulated contemporary with two-car garage
- Size: 2,600 square feet
- Price: $150,000 ($58 per square foot; excludes land and landscaping but includes site development)

Construction

- Oriented slightly west of true south
- 2x6 wall construction with 1-inch rigid insulation
- Continuous 6-mil vapor barrier
- All cracks and gaps sealed with polyurethane foam
- Overhangs sized for summertime shading
- Casement and awning windows on the east, west, and north of Heat Mirror insulated glass
- Lower level set back into the earth on the north side but open to the south
- Insulated metal door between home and garage is R-6; doors opening into the sunspace have R-2 metal doors

Insulation

- Cathedral ceilings: R-40
- Above-grade exterior walls: R-25
- Basement walls: R-22
- Under-basement walls: crawl space is sealed and insulated to R-13

Ventilation

- Operable casement and awning windows throughout
- Motorized awning windows at top of sunspace
- Natural and induced summertime ventilation

Heating and Hot Water

- Heating system: gas-fired boiler situated in garage, isolated from living spaces (AFUE: 80%)
- Distribution: Network of pipes runs under the plywood floor and is insulated (R-11) from underneath. The sunspace and the mechanical room in the garage are unheated
- Control: Three zones, each with its own thermostat
- Annual heating costs (with natural gas at $.34 per 100 cubic feet): $100-$200
- Hot water: Instant tankless wall-mounted heaters (gas)
- Annual hot water costs (with natural gas at $.34 per 100 cubic feet): $150

Special Features

- The garage has no mechanical heating, but is insulated and fitted with south-facing windows to collect solar heat.
- Wood-burning fireplace with tight doors and outside combustion air
- The angular nature of the home and room orientations to the sunspace give surprising and unusual views.

Architect's Comments

"The house is wrapped around a twenty-foot-high sunspace, or atrium, that buffers the living areas from the cold and provides free solar heat. Because of the buffering, heat loss from the living area is only about a third of what might be expected from a home this size, in this climate.

"As heat gathers at the top of the sunspace, it's collected in a central duct and blown down into the crawl space under the first floor. Registers allow the cooled air to return to the sunspace. During much of the heating season the sunspace is a net contributor of heat to the house; other times it's neutral. There are only about twenty days a year when the sunspace is actually losing heat and needs to be isolated from the house.

"All of the rooms have doors opening into the central sunspace and are designed to receive natural light. In our very dry climate, humidity inside the house stays nicely in the 40 to 50 percent range without any mechanical humidification."

— Rick Cowlishaw, architect

A Tale of Three Builders: *The Mingledorff Home*

Who, What, and Where

- Owner: Dr. Kurt Mingledorff
- Builder: Russell Home Builders, Pensacola, Florida
- Location: Pensacola, Florida
- Style: Superinsulated contemporary with two-car garage
- Size: 2,662 square feet
- Price: $198,500 ($74 per square foot; including land and site development)

Construction

- True south orientation
- 2x6 wall construction with 1-inch foam board and stucco
- Airtight construction with 6-mil vapor barrier
- Special trusses allow more insulation around top plate
- Overhangs adjusted for summertime shading
- Thermal breaks on all windows and doors
- All windows and doors foamed in place
- Low-E, argon-filled window and door glass

Insulation

- Ceiling: R-40
- Exterior walls: R-26
- Perimeter: R-7
- Radiant barrier in attic

Ventilation

- Operable casement windows throughout
- Heat recovery ventilator
- Exhaust only kitchen and bath vents
- Full soffit, ridge, and gable roof ventilation system
- Ceiling fans in all bedrooms and family room

Heating, Air Conditioning, and Hot Water

- Heating and cooling system: geothermal (ground-coupled) heat pump (EER: 16.2; COP: 3.78)
- Distribution: ducted forced air; all ductwork runs *inside* the conditioned envelope
- Control: zoned with programmable thermostats
- Hot water: Closed-loop solar hot water system with heat pump backup.

Special Features

- Fireplace: outside combustion air; two-speed blower; glass doors and damper
- Bacteriostatic water-treatment unit for drinking water and icemaker
- Electronic air cleaner

Builder's Comments

"Like many other Americans, I've become concerned about our nation's growing dependence on imported fossil fuels and the depletion of our natural resources. Over the past fourteen years I've researched many different construction techniques and energy-saving products. By culling the very best and refining them into an integrated whole, we developed the Energy Smart Home.

"The Mingledorff home, here in Pensacola, is representative of all our Energy Smart houses, designed and built to last, to be gentle on the environment, and to provide the owners with comfort, quiet, and value. I'm proud to report that the total cost for air conditioning, heating, hot water, ventilation, lights, and appliances in this home comes to less than $2.50 per day!"

— Philip Russell, builder

Forced Air Heating

If your house is reasonably tight and adequately insulated, yet you're still uncomfortable during the heating season or paying exorbitant fuel bills, it's time to take a long, hard look at your heating system.

Heating systems are most easily classified by the medium they use to distribute heat. Those that use air are called "forced air" or "warm air" systems. Regardless of what type central heater they have or the fuel they use, all forced air systems have ducts, made out of sheet metal or fiberboard, to move heated (or cooled) air around the house. Whether rectangular in shape, round, or a combination of both, the duct runs out of the central furnace or air handler and branches hither and yon to the various rooms of the house. An electric fan built into the central unit pushes heated air through the supply ducts to outlets in the floors, walls, or ceilings. Well-engineered systems, like the one shown on page 81, will have return air grilles and ducts to route cool air back to the central heater.

Home heating systems that use water or steam to conduct heat are lumped together in the "hydronic" category. Instead of ducts and dampers and air outlets, they have pipes and valves and radiators.

A third group, which I'll call "direct" for lack of a better word, has no distribution system. Electric baseboard radiators and wood stoves are good examples.

Out of the nation's 64.4 million single-family homes, about 41 million — or almost two out of every three — have forced air heating. Since forced air is by far and away the most common system around, it seems logical to discuss it first. "Hydronic" and "direct" systems will be discussed in chapters 7 and 8.

Those of you who live in Key West or Honolulu should feel free now to skip ahead to the chapters on ventilation and cooling. The rest of us will catch up with you later — maybe around the pool.

Why Maintenance Is Priority One

No one can imagine the full range of horrors that are hidden away in America's basements, garages, and utility rooms. I'm talking

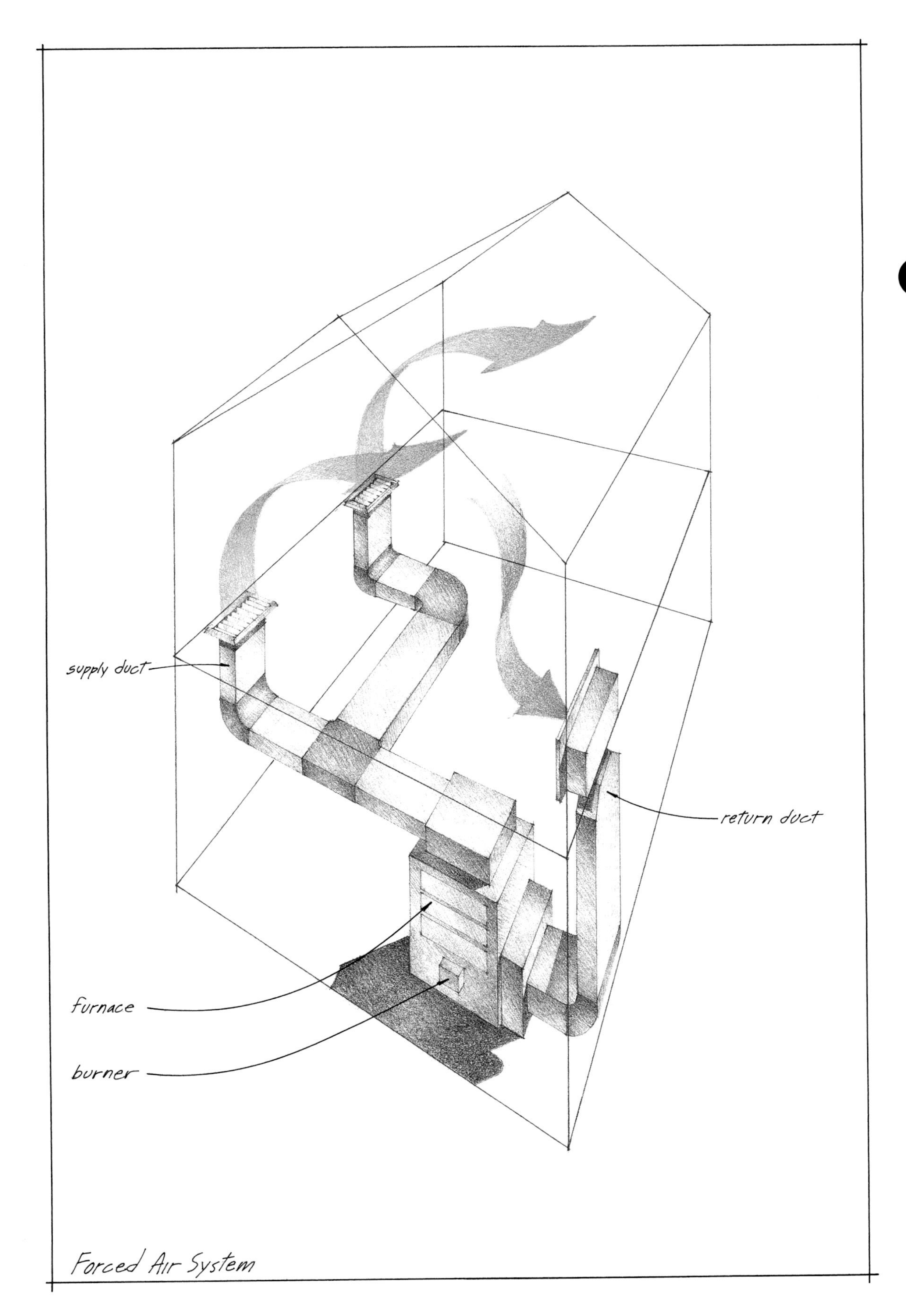
supply duct
return duct
furnace
burner
Forced Air System

What the Doctor Ordered

Recommended Maintenance for Forced Air Systems	Gas or Propane Furnace	Oil Furnace	Electric Furnace	Electric Heat Pump	Can I Do It Myself?
Clean the burner and combustion chamber; tune up the system	X	X			No
Run a combustion test to check furnace efficiency and preclude any danger of backdraft	X	X			No
Check for leaks in the gas line	X				No
Check the heat pump's efficiency[1]				X	No
Check for leaking refrigerant lines				X	No
Using appropriate pressure gauge and manufacturer's guide, charge the system with refrigerant				X	No
Test air flow	X	X	X	X	No
Clean or replace air filter	X	X	X	X	Yes[2]
Clean the fan	X	X	X	X	Yes
Change oil filter		X			No
Check fan belts for snugness and wear	X	X	X	X	Yes
Lubricate fan motors	X	X	X	X	Yes[2]
Adjust the fan switch	X	X	X	X	Yes[2]
Make sure registers are properly oriented and free of obstructions	X	X	X	X	Yes
Vacuum the ducts	X	X	X	X	Yes[3]
Check for duct leaks and make needed repairs	X	X	X	X	No[4]
Check, repair, and/or improve duct insulation	X	X	X	X	Yes
Clean the humidifier	X	X	X	X	Yes[2] [5]
Calibrate the thermostat	X	X	X	X	No[6]
Clean the indoor coil with a vacuum or brush	X[7]	X[7]	X[7]	X	Yes[8]
Clean the outdoor coil with a garden hose	X[7]	X[7]	X[7]	X	Yes
Keep outdoor heat exchanger coil free of tall grass and other obstructions	X[7]	X[7]	X[7]	X	Yes

See notes, opposite page.

Notes from chart, p. 82.
1. Test should determine its Energy Efficiency Ratio (EER) or Coefficient of Performance (COP).
2. Follow manufacturer's instructions.
3. Clean as far back into the registers as the vacuum will reach. For a more thorough job, you can hire a pro. But avoid firms that use gluelike materials to "seal" the dust inside the ducts. The sealers may contain formaldehyde or other harmful air pollutants.
4. Homeowners can check for "catastrophic" duct failures and patch failed duct tape, but it takes a pro with the right equipment to test and seal the system properly.
5. Many forced air systems don't have built-in humidifiers.
6. With an accurate thermometer and some patience you can detect a malfunctioning thermostat. But let your service rep clean or recalibrate it.
7. Only if the system has air conditioning.
8. If it's accessible.

about forced air heating systems that haven't been serviced for years; systems that waste 50 cents on every dollar that's paid out for fuel; systems that send particles of fiberglass or lead dust flying through the house every time the fan comes on; and — the most dreaded horror of all — systems that backdraft carbon monoxide and other combustion gases into the house.

A surprising number of HVAC comfort and economy problems stem from the simple fact that people are too lazy, forgetful, or inexperienced to maintain their systems properly. Most people wouldn't dream of driving their cars twenty or thirty thousand miles without an oil change, tune-up, and new air filter — yet that's exactly how they treat their heating systems. As you might expect, poor maintenance affects a furnace or heat pump the same way it does a car, with steadily declining performance, high operating costs, frequent breakdowns, safety problems, and finally, a premature trip to the scrap heap.

If you own a gas or electric furnace, I suggest you call in a *reputable* service technician every other year to have the system properly cleaned and tuned. For oil furnaces, which can get dirty faster, and heat pumps, which operate year-round, the system ought to be serviced *every* year.

Don't pinch pennies on the service call. Get someone who's good and pay them accordingly.

The checklist on page 82 details the key service elements for different kinds of forced air systems. When the service rep comes to call, be sure to tell him if you've had weatherization work done on your home or added any new combustion or ventilating appliances.

Many gas- and oil-fired furnaces draw their combustion air from *inside* the home. Tightening up the envelope on your house could diminish the amount of air available to the furnace for proper combustion and venting. So could installing a new clothes dryer, fireplace, bathroom exhaust vent, gas water heater, or gas range, which might compete with the furnace for indoor air. By running a combustion safety test on the furnace while the other appliances are on, the service rep can find out if there's any danger of backdraft.

Maintenance You Can Do Yourself

Even if you're not very handy, you can complement the professional service call by periodically doing some simple maintenance work yourself, as suggested in the list on page 82.

Simply by changing or cleaning the air filter at proper intervals you can keep the heat exchanger on your furnace or the indoor coils on your heat pump clean, which boosts performance, saves money, and extends the system's life.

A clean filter will also improve the quality of the air you breathe. The standard filter, which slides into a slot in the ductwork or blower compartment, is made of woven fiberglass or washable steel mesh. Simple and inexpensive as these filters are, they can remove plant spores, pollen, large-caliber dust, and some types of bacteria from the air. More sophisticated electronic air cleaners and extended media filters, which I'll discuss in chapter 10, go even further in protecting your lungs and HVAC equipment.

On the down side, air filters that are left unchanged for months and years at a time begin to choke off the flow of air through the system and become happy breeding grounds for bacteria and other bio-nasties. At some point, I reckon, you'd be better off to remove the filter altogether than to leave it in there clogged with dirt and bacteria.

Now that I've coaxed you into changing the air filter, why not go ahead and service the blower? To do this, you'll need to cut off electricity and fuel to the system. If you don't know how to do that, check the manual or ask your service rep.

Inside the air handler you'll see (perhaps with the aid of a flashlight) a cylindrical fan with curved blades, otherwise know as a "squirrel cage." (Engineers feel better if they call it a "centrifugal blower.")

If the fan blades are caked with dust and grime, scrape them off with a screwdriver or putty knife. Next, give the blower bearings a shot of oil so that the cage can turn freely. Finally, if it's a belt-driven assembly, check the belt for snugness, wear, and alignment, the same way you do with your car.

The blower motor probably won't need oil, since most are sealed and self-lubricating. But if you have an older style motor with an oil reservoir, it ought to be filled once a season, or as prescribed in the manual. (Don't overlubricate the motor — a couple of drops of lightweight oil is plenty.)

Since you've already got your hands dirty and your head stuck inside the air handler, why not clean off the heat pump or air-conditioner coils with a vacuum cleaner or brush? It will improve the system's performance, save money, and probably extend its life.

You may also want to adjust the fan switch, which is usually accessible on the outside of the cabinet. The switch is connected to a sensor that reads the temperature of the air coming out of the furnace. Its job is to optimize the fan's operation for comfort and economy.

For example, when the living room thermostat calls for heat and the furnace burner comes on, the fan switch cleverly waits until the air inside the furnace hits a certain comfortable temperature (the "on" set point that you chose) before it turns the blower on — otherwise you'd get a blast of uncomfortably cold air coming through the outlets. Conversely, when the thermostat sends a signal that its set point has been satisfied and shuts the burner off, the fan switch realizes that there's still a lot of heat inside the furnace that can be harvested. So it continues to blow for a few minutes until the air temperature falls back below the comfort level (the "off" set point you chose).

Some service reps are in the habit of adjusting the fan switch to an "on" setting of 135°F and an "off" setting of 100°F so that they can be absolutely, positively, 100 percent sure that they'll never ever get a complaint. In other words, they err on the side of guaranteed comfort at the expense of economy.

To frugal-minded homeowners, I might suggest an "on" setting of 110°F and a shut-off set point ("off") no higher than 90°F. If there's only one setting on your system, try it at 110°F. Of course, every house, climate, and heating system is different, and no two people have the same definition of comfort. You may want to experiment with the fan switch to find a good balance point between comfort and economy. If you can't find the switch, ask your service rep to show you where it is.

My sympathies are with you if you're unable to service or adjust some of these components because you can't get to them or because the service manual is written in techno-jargon. Some of the better HVAC manufacturers are working hard to make their systems more accessible and easier to service. Some even include manuals that are written in plain English.

A Big-Time Offender

One of the most notorious sources of comfort and cost problems — on all kinds of heating systems — is a faulty thermostat. While thermostats rarely fail outright, they can degrade over time as their calibration slips and mechanical parts begin to stick.

A broken thermostat can produce wild swings in temperature as it delays turning the system on or fails to turn it off. The result, of course, is uncomfortable people. And when people get uncomfortable, they tend to crank the thermostat way up or down, which only makes things worse.

Contrary to popular belief, thermostats don't work like gas pedals. When you jam the thermostat's set point way up because you're cold, the air delivered to the room won't come out any faster or warmer than it would if the thermostat were set only a degree or two above room temperature. Unlike a car engine, which revs up when you push the pedal, conventional furnaces and heat pumps

The Brain's the Thing

One of the most cost-effective improvements you can make to an old HVAC system is to change its brain for a new one. By installing a new setback thermostat (sometimes called a "clock" or "electronic" thermostat), you'll be able to automatically raise or lower indoor temperature to fit your schedule.

During the winter, for example, you could program the thermostat so that it lets indoor temperature settle back to 60°F while the family sleeps, then automatically brings the temperature back up to 70°F before everyone gets up. A second 10-degree setback might start when the family leaves for work and school. If there's no one home after 8:00 A.M., for example, the thermostat could give the furnace a rest, letting the indoor temperature slide back to 60°F again. At 3:00 P.M., just before the kids get home from school, the house would be automatically warmed again. If there's ever a change in your schedule — you stay home from the office, for example, or Aunt Ida comes for a visit — you can simply override the program.

Without making any sacrifices in comfort, you can enjoy a 10 to 20 percent savings on your annual heating bill — simply by not heating the house when you don't need it. In the summertime, the thermostat would help save on air conditioning costs too — just by letting the house warm up a little when there's no one at home.

Setback thermostats cost from $15 to $160, but I'd steer clear of the ultra-cheap ones. Who wants a ticky-tack thermostat running thousands of dollars' worth of heating and cooling gear, especially when you're not at home?

How much money you save with a clock thermostat will depend on your climate, local fuel prices, the airtightness and insulation values of your home, and the efficiency of the HVAC gear. No matter what the variables, a new setback thermostat will probably pay for itself in saved energy in less than two years. After that, it's money in the bank.

The most flexible (and expensive) thermostats on the market have seven-day programs, that is, they can hold a different program for every day of the week. Somewhat less flexible are the 5 + 1 + 1 thermostats, which have a single Monday-through-Friday program, with separate programs available for Saturday and Sunday. Less flexible still are the 5 + 2 models, which hold just two programs, one for the five workdays and one for the weekend. Last, and least flexible of all, are the one-program models.

Your lifestyle should be your guide in selecting the type of thermostat you need. Don't pay for more flexibility than you can use. If all you want is a simple night setback thermostat that repeats its program seven days a week, there's no need to invest in a fancy programmable model. A simple and inexpensive thermostat with mechanically adjusted setpoints will do just fine.

If you do opt for a programmable model, here are some features to shop for:

- Easy programming: Though your instincts may tell you that fewer buttons are better, the inverse is usually true. Models that skimp on keys have to assign each key a multiple function to get the job done. Single-function keys tend to make programming easier. If you can program a VCR or microwave without panicking, you should be able to handle a programmable setback thermostat. (Otherwise, you'd be better off getting an electromechanical model that uses small levers or pins to mark the setbacks.)
- Manual override: This lets you manually override the current temperature setpoint without reprogramming the thermostat. I recommend models that automatically revert to the original

are either "on" — if the thermostat's set point isn't satisfied — or "off" — when it is.

What you'll probably succeed in doing, as you settle down in front of the television set or report to the kitchen to cook dinner, is forget all about your rash little act with the thermostat. But the thermostat will remember, and conspire with the furnace to raise the indoor temperature to 90°F — just the way *you* wanted it. This leads to the next stage, which finds you storming into the living room, sweating profusely, to whack the set point down to 50°. Which then leads to the next stage . . .

program after the override period has ended, so that you don't end up leaving the thermostat in its override mode.

- Easy changeover from heating to cooling: Most electronic thermostats have a key or switch that lets you easily change from a heating season program to cooling, or vice versa, without reprogramming. State-of-the-art models will make the changeover for you automatically, a nice feature if you live in a climate that sometimes calls for heating and cooling in a single twenty-four-hour period.
- Low-battery indicator: Most electronic thermostats use batteries to hold the program in case of power failure, but some use batteries as their sole source of power. Thermostats without a low-battery indicator can fail without notice — maybe when no one is home.

If you plan to install a new thermostat yourself, the first step is to make sure you buy one that's compatible with your equipment. One way to do this is to remove the old thermostat from the wall and count the number of wires. (*Make sure the electricity is off before you do this!*) Heating-only and cooling-only systems have just two or three wires, which makes them a snap to hook up. Combined heating-cooling systems typically have four or five wires, while heat pumps use eight or nine.

If yours is a 24-volt system (most are) and you've done some simple electrical wiring jobs before, you should be able to install the thermostat yourself. However, if you have a 120-volt or millivolt system, I'd suggest you call in a pro. A pro would also be recommended for zoned systems, pulse furnaces, and heat pumps, because of their added complexity.

If you do install the unit yourself, here are some tips:

- Mount the thermostat on an interior wall about 5 feet off the floor.
- Don't use a wall that's facing unconditioned air space or that has pipes or flues in it.
- Avoid putting the unit near air outlets, radiators, or drafty spots.
- Don't place the thermostat in a closet or behind a door.
- Make sure the spot you choose doesn't get direct sunlight.

Good as today's electronic thermostats are, I think we're going to see even better ones in the years to come. One key improvement is to equip systems with more and better sensors so that the central controller can get the feedback it needs to make smart adjustments.

"This Old House" may have been the first TV show ever to demonstrate how an outdoor-temperature sensor can improve the sensitivity and economy of residential HVAC controls. In fact, we put one in on our very first project, back in 1980.

The really smart thermostats of the future will probably be able to sense not only outdoor-air temperature, but relative humidity and wind conditions as well. It is climate, after all, that defines a home's heating and cooling needs — without current weather information, a thermostat is, in a manner of speaking, half blind and mostly deaf.

As the concept of zoning becomes more and more popular, control systems will have several subordinate thermostats and/or remote sensors throughout the house, so that the central controller has real-time information to work with. Instead of heating or cooling all of the rooms in a house all of the time, smart thermostats will focus on heating and cooling *people*, where they need it, when they need it.

Some of the newer HVAC systems can vary the rate at which the burner (in furnaces) or compressor (in heat pumps) runs. This, combined with variable-speed fan motors and more sophisticated controls, enables these systems to respond with greater subtlety and flexibility. We'll talk more about these improvements when we discuss buying new equipment.

With a little time, patience, and an accurate room thermometer, you can test your thermostat to see if it's turning the system on and off as it should be. (Note the cycling of the furnace or heat pump as your cue, *not* the blower.) If you suspect you have a problem, ask your service rep to check it out. Cleaning and recalibrating an old thermostat can sometimes solve the problem.

Or maybe it's time to go shopping for a new one. Replacing a worn-out thermostat could be one of the most cost-effective and comfort-enhancing investments you'll ever make, and could add a whole new life to your old furnace or heat pump.

As the sidebar on page 97 explains, you can buy a new thermostat for your system for as little as $40.

Poor Ductmanship

Without a doubt, one of the most common and serious problems that affect forced air heating systems is poor-quality ductwork.

Jim Cummings, a researcher at the Florida Solar Energy Center, checked out some typical Florida homes — 150 to be exact — to see how well their heating and air conditioning ducts were working. After finding that the average home lost 11 percent of its heat through leaky ducts, he concluded that improving the quality of ducts is "the single biggest opportunity we have to save energy in the United States."

When Jim and his team went back into those homes and properly sealed and insulated the ducts, the average heating and air conditioning bill fell $110 a year. Though it wasn't documented, I'm sure there was a parallel improvement in homeowner comfort.

With more than a million miles of heating and cooling duct installed in the United States, and much of it leaking like a sieve, we face both a huge problem and a golden opportunity. By plugging those leaks, homeowners can actually get *more* comfort for *less* money.

From a broader economic and environmental perspective, fixing leaky ducts — if enough homeowners get serious about it — would take pressure off the utilities to build new generating plants and pipelines, which would help keep energy costs down, conserve natural resources, and avoid new sources of pollution. Jim Cummings estimates that if duct tightening and insulating were done statewide in Florida, it would save 4,500 megawatts of peak generating power — 13 percent of the state's present capacity. No wonder Florida Power & Light, the state's largest utility, is getting behind the repair work in a big way.

Of course Florida isn't alone in this. Joe Iorio, who heads Atlantic Heating and Air Conditioning Co., based in Brookline, Massachusetts, says that only about half of all the forced air systems installed in New England are properly ducted. Inspectors in California, North Carolina, and other states say the quality of the ductwork there is just as bad. Maybe worse.

Sealing and insulating the ductwork on forced air systems helps insure that the heat arrives at its final destination.

The problem has two basic sources. First, a lot of residential ductwork was slapped together with no real effort to seal it properly in the first place. Second, systems that *were* properly sealed are starting to leak as the duct tape — exposed to heat, humidity, and air pressure — starts to crack and peel away.

In homes where the ductwork runs through conditioned spaces — that is, the lived-in areas of the home that are heated and cooled — this may not create too big a problem. Unfortunately, most ductwork runs through attics, crawl spaces, garages, and unheated basements. So when the blower comes on and pressurizes the ductwork with heated (or cooled) air, much of that air never arrives at its destination.

The biggest problem areas tend to be:

- leaks in the return air plenum;
- leaks where branch ducts are attached to the main trunk line;
- and leaks where the duct is attached to a room outlet.

Flashbacks: The Octopus in the Basement

Before the introduction of electric fans put the "forced" into warm air systems, they operated entirely on the principle of convection — that is, warm air rising (because it's lighter) and cold air falling (because it's heavier).

Most of the original "gravity air furnaces," like the one shown here, were coal- or wood-fired, with very large diameter ducts sweeping out of the furnace and back. Coming across one of these in a dark basement is like confronting a giant octopus, tentacles spread, ready to seize you by the throat. (What really happens, of course, is that you bang your head against a low-hanging duct.)

The large ducts, with their gently curved elbows, were designed to create as little friction as possible so that warm air could rise and cold air fall on very small differences in temperature and pressure. Some of these gentle giants are still in place, but most have been converted to oil or gas now and fitted with a not-so-gentle blower.

If you're going to change the furnace on one of these old-timers, I'd strongly recommend replacing the ductwork too. You can't drop a Maserati engine into a 1958 DeSoto and expect very good results.

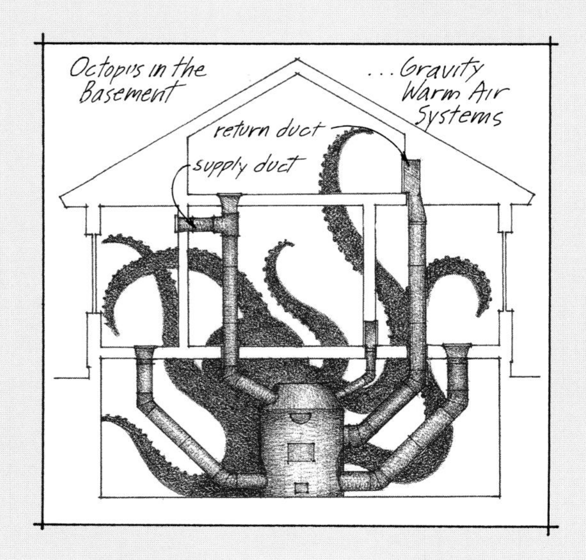

In addition to numerous small- and medium-size leaks, house sleuths routinely find "catastrophic" duct leaks — that is, instances where a section of duct was never installed or connected, or has broken clean away, leaving a gaping hole in the system.

Apart from the wicked effect this type of failure has on the homeowner's comfort and economy, it also jeopardizes the family's health. Imagine a piece of broken duct that's open to the attic, sucking in bits of fiberglass or cellulose insulation, which ends up wafting out into your bedrooms and kitchen. This is no way to win awards from the American Lung Association, I can tell you. Or, even more frightening, picture a duct running through a crawl space, cracked at the seam, sucking up a fresh layer of termite poison or radon soil gas.

The best way to resolve these kinds of comfort, cost, and safety problems is to have a savvy HVAC contractor run a flow hood or blower door test on your ductwork. Both tests are effective in revealing the total amount of leakage in the system and pinpointing where those leaks are. A gooey mastic is then used to seal the leaks. (Because of its poor history, duct tape is no longer recommended for sealing or resealing ductwork, and has already been outlawed in some building code jurisdictions.)

I should warn you ahead of time that you may have real trouble finding a contractor who's up to speed on this. Only in the last year or so have the national laboratories, testing and standards organizations, trade associations, and utilities realized how serious the problem is and begun to train people to address it. If you do find someone who knows how to test and repair your ductwork, it will probably set you back $200 to $500, depending on the size of your house. That's not a bad investment, though — you'll get more comfort, less frustration, and a two- to three-year payback on your money. After that, it's pure gravy.

Though you may be tempted to seal or reseal the ductwork yourself, it's probably not worth the effort, and could even make things worse. Without the proper test gear, even HVAC pros have no way of knowing how much the system is leaking and where.

As an interim measure, I think you *should* check the system for catastrophic duct failures — the telltale symptoms, as I mentioned before, are gross discomfort, sky-high utility bills, and foreign matter or odors coming out of the registers.

Another *doable* do-it-yourself job is to make sure that any ducts running through unconditioned spaces are wrapped with an R-11 blanket of fiberglass. The insulation won't stop air leaks, of course, but it will help slow conductive heat losses. If you're going to have the ductwork pressure-tested and sealed by a professional, wait until that work is completed before you insulate the duct so that you won't have to do it twice.

While you're making your inspection, check the seams around any duct that penetrates an exterior wall, floor, or ceiling. If there's a gap that isn't properly sealed, arm yourself with a caulk gun and take care of it.

Complaints about "Dry" Heat

As I travel around the country, the most common complaint I hear about forced air heating systems is that they "bake" or "dry out" the air, which makes people feel uncomfortable.

But the truth is, forced air furnaces don't remove *any* moisture from the air. What's actually happening — and the source of all these complaints — is quite a bit more complex.

First of all, forced air systems are often out of balance — that is, the supply of warm air being blown into a room isn't in equilibrium with the cool air flowing out. So some parts of the house are pressurized when the fan comes on, while others are depressurized.

Imagine, for example, a bedroom that has a warm air supply outlet, but no return grille. (The cold air return might be located in the hall, serving several rooms or perhaps the whole house.) When you close the bedroom door, you also block off the only pathway the return air has to the furnace. So what you've done, in effect, is pressurize the bedroom. Meanwhile, with the system thrown out of balance, other parts of the house are depressurized. And since nature abhors a vacuum, she pushes outside air — cold, *dry* air — in through the walls. As dry winter air invades the house and the relative humidity falls, people start to feel cold and dry. And of course the furnace has to work harder too.

If the ductwork is leaking into the attic, crawl space, or garage, the depressurization — and dryness — is even worse. Every time the blower comes on, it sucks air out of the house and pumps it outside, while cold, dry air filters in elsewhere to replace it. (If the ductwork is used in the summer for central air conditioning, the resulting depressurization pulls hot, humid air into the house.)

A recent survey published by the Electric Power Research Institute found that infiltration rates are 15 to 36 percent higher in homes with blowers — that is, forced air systems — compared to houses without them. This lays bare a fundamental weakness in the way forced air systems are being built and documents serious problems with their comfort and economy.

I should point out that even if the supply and return air on the system are well balanced and the ductwork is tight, the furnace itself — if it's drawing its combustion air from indoors — can depressurize the house.

To be honest, I don't think there's a perfect fix to these problems. But here are some tactics that can help address the dryness and also reduce your fuel bills.

- Raise the relative humidity inside your home by adding house plants or buying a portable humidifier. This is a superficial solution, to be sure, but it'll help make you feel more comfortable.
- Add a permanent, built-in humidifier to the air handler. You're still treating the symptom rather than the cause, but now have the convenience of an automatic humidifying system that treats the whole house.
- Make sure your ductwork is well sealed and insulated, and that none of the ducts is blocked. This can be a real source of problems.
- Ask your service rep how to put the system in better balance. In addition to adjusting the dampers, he may recommend leaving certain interior doors open, or trimming their bottoms off so that a return air path remains even when they're closed. Another tack is to put in new supply

or return air ducts, or through-the-wall vents to balance two adjoining rooms, but this type of retrofit starts to get expensive and disruptive. (Don't tinker with the dampers yourself or try to regulate the system by opening and closing air outlets in the rooms. This is liable to throw the system even further out of balance.)

- Tighten up your house using the weatherstripping and caulking techniques I discussed in chapter 4. Tightening the envelope will slow the passage of cold, dry air from outdoors to in and raise the relative humidity level inside. You'll feel warmer when the relative humidity goes up, even if the air temperature hasn't changed.
- Seal off the furnace from the rest of the house, preferably in a small room that's weatherstripped and insulated, and provide it with a vent to bring in outdoor combustion air. This stops the furnace from depressurizing the rest of the house and eliminates the danger of backdraft. It also saves energy, since the furnace is no longer using previously heated air for combustion.

When Warm Feels Cold

Forced air heating has a second comfort disadvantage that causes lots of wintertime complaints. Despite the fact that the air blowing out of the register is 90°F or warmer, the air moving over people's skin can leave them feeling cool and dry.

Whether it's wind sweeping along the side of a building or a child blowing on his hot soup, moving air always promotes convective heat losses (cooling) from the surface beneath it. In this case, the moving air also increases evaporation off our skin, which means more heat lost and a colder feeling still.

The discomfort caused by moving air is especially pronounced with heat pumps, which typically deliver air at 90°–110°F. Because most oil and gas furnaces produce outlet air temperatures that are fifteen or twenty degrees higher, the effect isn't as noticeable.

Since warm air can hold more moisture than cold, the air coming out of the register also has a drying effect as it sweeps across the room. Imagine, if you will, a big dry sponge pushing out of the register every time the fan kicks on, producing a distinct and fairly rapid drop in the relative humidity.

Other types of heating systems — such as baseboard hot water — also produce this localized drying effect, but with less discomfort because of the gradual way in which heated air is blended into the room.

One way to address the problem of air movement in warm air systems is to equip the supply registers with diffusers or diverters that diminish or redirect the airflow away from your dinner table or favorite easy chair. If an adaptor won't solve the problem, you may have to have your HVAC contractor install a different-size supply register and/or move it elsewhere in the room.

Of course, the best way to have a draft-free system is to have it properly designed and installed in the first place. To do that, the contractor has to consider not only the capacity of the central heating unit and the duct layout, but also to pay careful attention to the type, size, and location of the supply and return registers.

The Power of Zoning

If you've followed our HVAC work on "This Old House" through the years, you know me to be a strong advocate of zoning. By dividing a house into two or three different zones — each controlled with its own thermostat — you can deliver a lot more comfort for a lot less money. Zoning does away with "thermostat wars" between hot- and cold-blooded relatives. And because heat is delivered only to those rooms where it's needed, when it's needed, waste is kept to a minimum.

Until very recently, though, I've had to limit my enthusiasm for zoning to homes that have hydronic or electric baseboard heating. Efforts to zone forced air systems, I knew, had produced discouraging results in the industry, including systems that were finicky to balance and control, prone to breakdowns, and frequently noisy. The only proven way to zone a forced air system was to double up the equipment, using, for example, one heat pump for the ground floor and another, with a separate thermostat and ductwork, for the second floor, which is, of course, very expensive.

But the world has really changed in the past couple years. The advent of more powerful, less expensive microprocessor controls and motorized dampers that can reliably modulate airflow has made forced air zoning practical at a reasonable cost.

Carrier's new Comfort Zone, designed specifically for forced air systems that have a *single* furnace or heat pump, makes cost-effective zoning possible in homes as small as 1,400 square feet. The system enhances indoor comfort by creating up to four zones, each with its own thermostat. Sensors located in the ductwork continually report on the system's status so that temperature, humidity, and airflow adjustments can be made. The system also has an outdoor sensor so that it can react to changing weather.

Comfort Zone is most easily installed in new construction, but can also be retrofitted, so long as the existing ductwork is

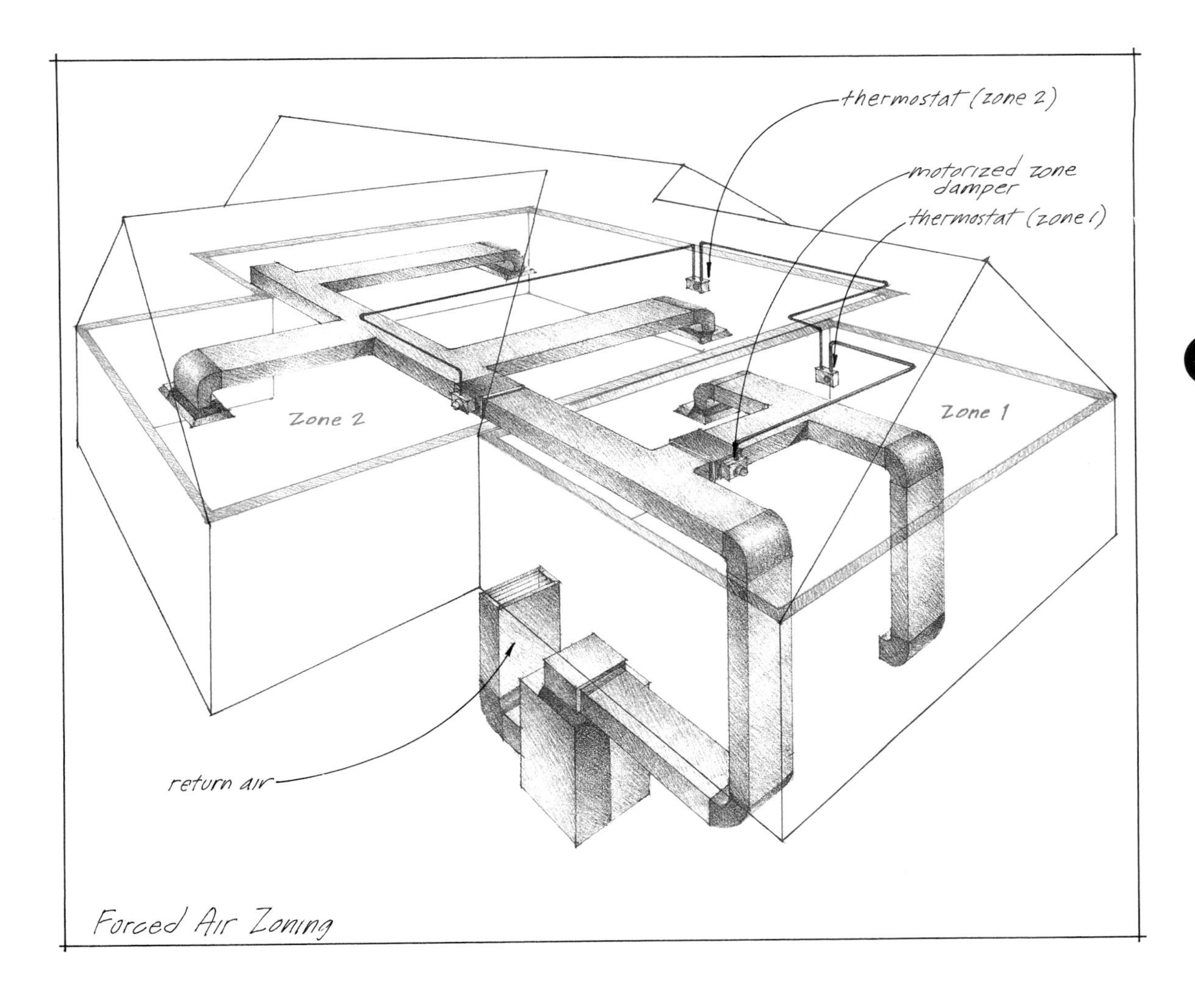

adequately sized. Priced under $2,000, the system should pay for itself in saved energy in as little as three and a half years, the company says.

The Harmony forced air zoning system, from Lennox, is also capable of handling four zones and priced to yield a three- to five-year payback. It comes packaged with Lennox's new variable-speed Pulse gas furnace.

On another front, Honeywell has teamed up with Trol-A-Temp, Inc., a pioneer in home zoning, to create a new forced air zoning system of its own. The company says the Honeywell/Trol-A-Temp system can handle up to three zones, with an installed cost of about $300 per zone in new construction and $400 to $500 per zone in retrofits.

I'm excited by the advances that have been made in forced air zoning systems and their availability now as a viable retrofit option. If your existing control and distribution system is beyond the point of patching, this could be a terrific fix.

Profitable Patches: For Forced Air Furnaces

Item[1]	Cost	Estimated Fuel Savings[2]	Comments
Replace fuel nozzle with a smaller one	$0–$60[3]	2–10%	Easier and less expensive with oil furnaces than with gas.
Flame-retention head oil burner[4]	$250–$600	10–25%	Oil furnaces only. More effective than downsizing the nozzle, but also more expensive.
Replace pilot light with electric ignition	$150–$300	5–10%	Gas furnaces only.
Install automatic vent damper	$250–$400	3–15%	Closes off the flue when furnace quits firing to reduce stand-by heat loss up the stack. Savings are usually greater with oil furnace than with gas.
Install a clock thermostat	$40–$280	7–10%	Enables you to lower the temperature automatically while you're sleeping or away from home.[5]
Install gas power burner	$400–$600	10–20%	For converting old oil and coal systems to gas.

1. Consult a good HVAC contractor or energy auditor about the cost-effectiveness of modifying your furnace versus buying a new one.
2. Savings on these items are *not* cumulative.
3. On oil furnaces (but not gas), the nozzle can often be downsized for free as part of a tune-up.
4. May require downsizing the combustion chamber.
5. Do-it-yourself installation is possible on some heating and cooling systems. See page 87.

Evaluating Your Central Heating Unit

If you've maintained your system well and don't have any serious thermostat or duct problems, the only place left to cast a suspicious eye is on the furnace or heat pump itself.

Trying to decide whether to upgrade the unit or replace it with a new one can be an agonizing decision. Generally speaking, if a furnace is more than 15 years old or a heat pump more than 10, or if the unit was radically oversized to begin with (a very common practice), you should probably go ahead and junk it.

If the system is younger than that, or you don't have the $1,500–$4,000 that it would take to buy a new furnace or heat pump, take a look at some of the fix-it-up measures I've laid out in the adjoining chart.

Assuming that your current system is 55 or 60 percent efficient in converting its fuel to usable heat, a patchwork solution might nudge it up to 70 or 75 percent. I urge you to take full advantage of your HVAC contractor's experience and integrity in putting your money where it will do the most good.

In some cases, of course, it's a lot smarter to buy new than to sink any more money into a junker. If you do decide to buy new, I must tell you right from the start that there's no such thing as *the* perfect system. In my many years as a contractor and member of the "This Old House" team, I've learned that every system has its good points and bad. It's in knowing those points and weighing the tradeoffs that you make your best decision.

Oil and Gas Furnaces

New furnaces are best categorized by their efficiency — that is, how much usable energy they can deliver from each unit of fuel. The U.S. Department of Energy tests and rates oil and gas furnaces, assigning each model an Annual Fuel Utilization Efficiency, or AFUE. While there are other ways to measure furnace performance (by combustion efficiency, for example), AFUE is the most realistic.

As your contractor will explain to you, AFUEs on new furnaces range from 78 percent efficiency on the low end all the way up to the high nineties. Furnaces with efficiencies above 90 percent are called "high-efficiency" or "condensing" furnaces.

The installed cost of a high-efficiency furnace is typically 25 to 30 percent more than a conventional model. But since it burns a lot less fuel, the high-efficiency model could turn out to be the smarter investment over time, especially in larger homes with sizable heating bills.

Why the extra up-front costs? In part it's because high-efficiency furnaces come equipped with two separate heat exchangers. Hot combustion gases that would ordinarily go shooting up the flue are forced through an additional loop of pipes (the extra heat exchanger) where they surrender more and more heat. Finally, the hot gas cools to the point where water vapor condenses out of it — hence the name, "condensing" furnace.

Not only are the heat exchangers in condensing furnaces larger, they're also built out of special types of stainless steel and other alloys that can stand the hot, corrosive environment that's created inside the furnace. These higher-grade materials, along with special welding techniques and coatings, also add to the cost.

Operating at full tilt, a condensing furnace will produce several gallons of water a day, which is drained out of the heat exchanger into the home's sewer line. With a pH ranging from 4 to 2.4, the acidity of the condensate coming out of the system is comparable to Coke or lemon juice. Though it's usually not necessary to protect sewage pipes, some installers equip the drain line with a small, inexpensive, lime-filled cartridge to neutralize the acid.

While the water draining out the bottom of a condensing furnace won't cause problems, the cooled combustion gases going up the flue could. Imagine an old chimney, if you will, designed to han-

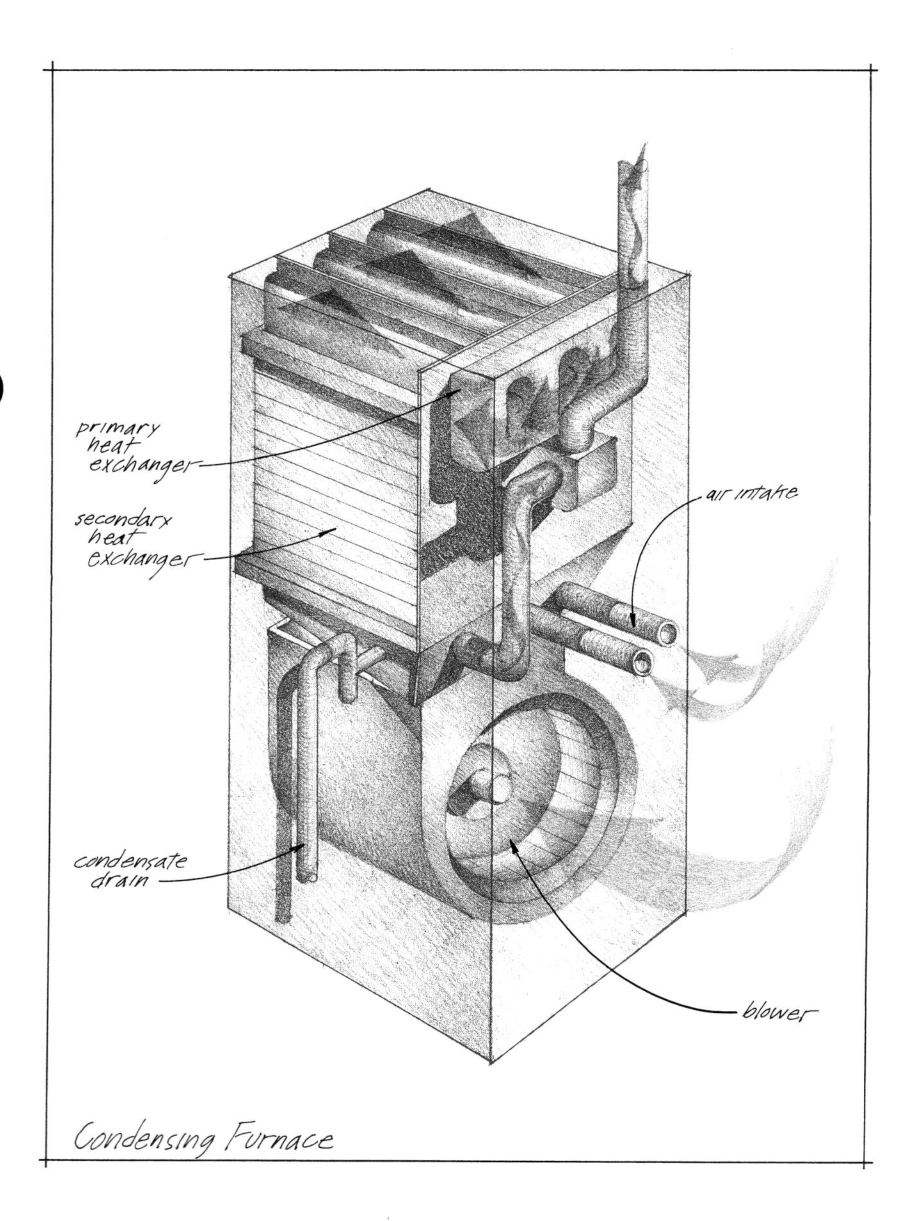

Condensing Furnace

dle smoke and gases blowing out of the furnace at 300°F or 400°F. Mixed in with those hot gases is a lot of water, of course, but it's harmless as long as it stays in vapor form.

So one fine day you change the old furnace for a high-efficiency model, and suddenly the gases coming out of the furnace are down around 100°F. Instead of the water vapor pushing on out the top of the chimney, it cools and condenses inside, depositing a potent brew of acid on the chimney liner and mortar.

Frankly, it scares me when I think how many contractors have slapped in new heating systems without really examining the flue.

So Tip Number 1 is: Whenever you buy a new furnace, whether it's a condensing model or not, make sure that your flue is going to accommodate it safely. Beware of contractors who disregard needed work on the flue so that they can low-ball the price.

One way to fortify an old chimney is to insert a new stainless steel or aluminum liner. Or you can use an inflatable form and pour in a new concrete liner, the way we did at the Weatherbee Farm in Westwood, Massachusetts.

One nice thing about condensing furnaces is that the flue gas temperature is so low — around 100°F — you can use a special type of plastic pipe for the flue. If building codes permit, your contractor can simply run a stretch of this pipe up the existing chimney. If that's not permitted, you may want to abandon the chimney altogether and run the plastic flue pipe out through the wall.

Being able to use plastic flue pipe in new construction is a real cost-cutter. The money you save in *not* having to build a masonry chimney can more than pay for the extra cost of the high-efficiency furnace.

Isolating the Fire from the House

Perhaps the best thing about high-efficiency furnaces (and some of the better lower-efficiency models as well) is that most of them no longer draw their combustion air from the house. By equipping the furnace with its own fresh air vent from the outside, and drawing air through the combustion chamber with small fans, state-of-the-art furnaces are completely isolated from the house. No more danger of backdraft. No more sucking already heated air out of the house for combustion.

Isolating the furnace also eliminates at least one source of depressurization, so there's less cold, dry air pushing its way into your home. And the furnace's heat exchanger will last longer too, because damaging indoor air pollutants (like the fumes coming off chlorine bleach and other types of cleaners and aerosols) are kept out of the combustion air. When those kind of airborne pollutants get into a condensing furnace, they form hydrochloric and hydrofluoric acid, which can speedily chew holes in a heat exchanger.

So Tip Number 2 is: Opt for a closed-combustion furnace. I can't think of an improvement that has done more to advance furnace safety, performance, economy, and longevity. In fact, I think closed combustion will become standard for *all* furnaces within the next few years. (If you can't afford this feature right now, the next best thing is to isolate the furnace in a small room, as I described above, and run in a makeshift vent from outdoors.)

Tip Number 3 goes hand-in-glove with Tip Number 2: Be alert to the fact that installing a new furnace can alter your home's ventilation needs. Taking out an old aspirating furnace and replac-

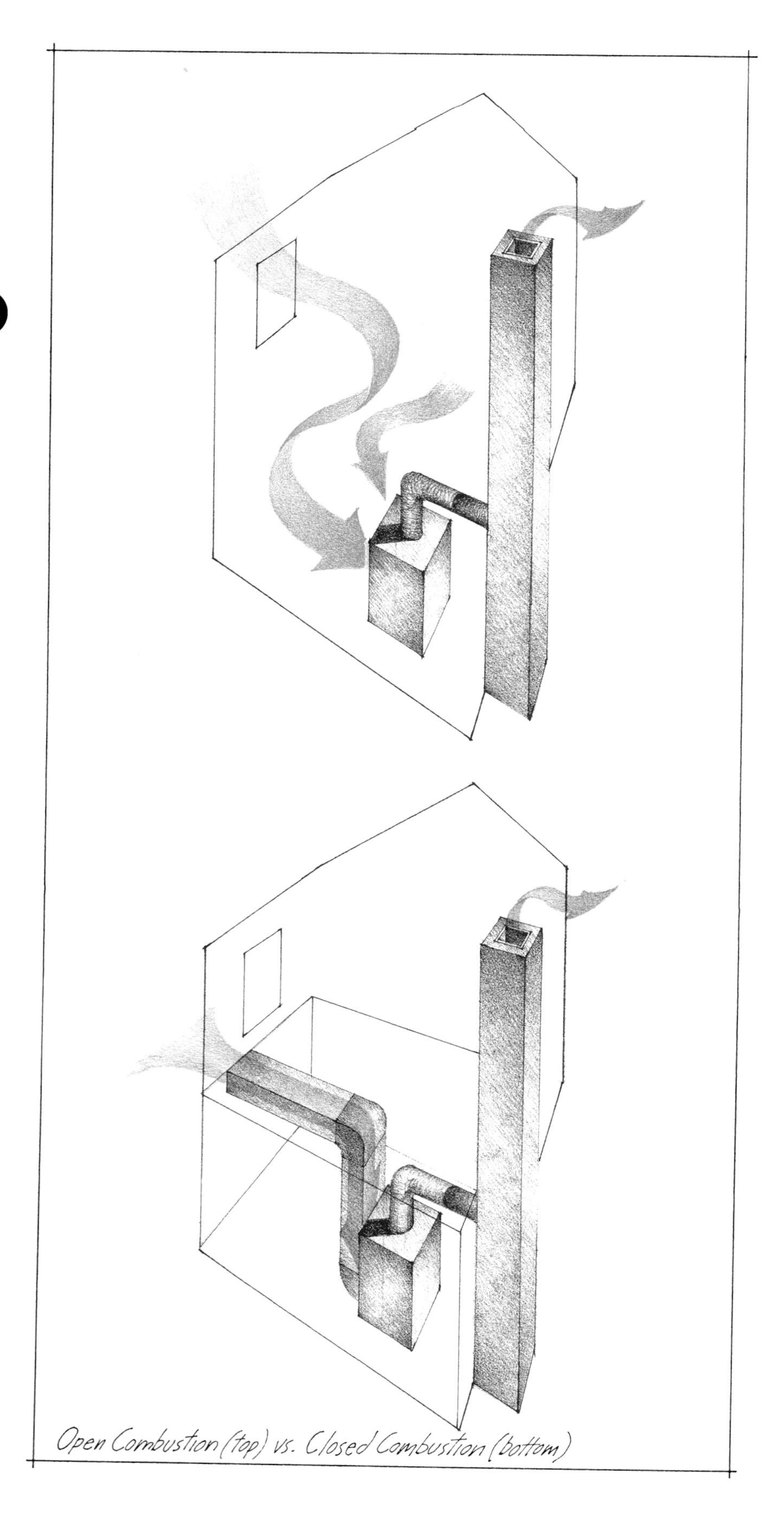

Open Combustion (top) vs. Closed Combustion (bottom)

ing it with a closed-combustion model removes the coincidental ventilation the old furnace supplied. To avoid indoor air-quality or moisture problems, you may need to install an exhaust fan, heat-recovery ventilator, or other controlled ventilation. In any case, be sure to discuss the issue with your contractor. (I'll have more to say about ventilation in chapter 10).

Tip Number 4: Assess the *real* value of efficiency before you buy. Despite all the hype, it doesn't always pay to buy the most efficient equipment on the market. Paying $1,000 extra for a high-efficiency gas furnace may make good sense in Minneapolis, where the length and severity of the winters could provide a quick payback. But the same investment in a Dallas home — where the heating season is relatively short — could turn out to be a flop. (Inversely, the payback would probably favor Dallas if we were talking about high-efficiency air conditioning.) The length of the season, fuel prices, and the size and condition of your home should all play a part in deciding how much efficiency you want.

Finally, here is a list of features that you ought to consider before you buy a new forced air system:

- Variable-speed blower: Gives the system the ability to come on slowly and quietly, ramping the fan speed up or down as demand calls.
- Built-in humidifier: Combats the dryness problems we discussed earlier.
- Built-in electronic air cleaner: At $500 to $700, this is an expensive option, but it extends the life of your lungs as well as the life of your HVAC gear.
- A strong warranty: For oil and gas furnaces, look for a *minimum* 20-year warranty on the heat exchanger. (Some manufacturers offer a "lifetime" warranty, meaning the life of the furnace.) For heat pumps, I'd want at least a 5-year warranty on the compressor. (Some manufacturers guarantee it for 10.)
- A reputable manufacturer: Lennox, Carrier, Trane, and Amana all make good-quality gas-fired furnaces and electric heat pumps, and have established long track records in the industry. If you want an oil-fired furnace, Armstrong, DMO Industries, and Hallmark are a couple names you can trust.

Heat Pumps

Choosing a new heat pump for your forced air system can give you some striking advantages — and some drawbacks as well. Because heat pumps use a reversible refrigeration cycle, they can produce either hot or cool air. No other system can boast of meeting all your heating and air conditioning needs with one fairly compact unit. (As we'll see in chapters to come, some heat pumps also provide hot water, so you actually get a 3-in-1 combination.) Another plus for heat pumps is that they're all-electric, so they don't need a flue or chimney. Nor do they need a gas line, propane tank, or fuel oil deliveries — electricity is already at the site.

Because heat pumps perform both as heaters and air conditioners, the Department of Energy assigns them two separate ratings. The first, called the Seasonal Energy Efficiency Ratio, or SEER, is a measure of cooling. From worst to best, heat pumps currently on the market have SEERs running from 10 to 17. The second rating is called the Heating Season Performance Factor, or HSPF, which is a measure of heating efficiency. These range from a miserable 6 to a masterful 10.

Since most heat pumps have been installed in the Sunbelt, with air conditioning as their primary task, I'll cover them in detail in an upcoming chapter on cooling. For now, suffice it to say that air-to-air heat pumps are not the ideal heating machine for New England or other cold-weather climates.

Why? Because as air temperatures fall toward freezing and below, air-to-air heat pumps have to rely more and more on their back-up heater, which is usually an electric resistance coil built into the air handler. I don't know how things are where you live, but here in Boston, where electricity costs 12 cents a kilowatt hour, homeowners who want to stay solvent shy away from electric resistance heating.

It's important not to confuse air-to-air heat pumps with the ground-coupled variety, which is the type we installed in the "This Old House" Brookline project. Ground-coupled heat pumps *are* appropriate for New England and other cold climates. (I'll have more to say about them in chapter 11.)

One final point about heat pumps is that if you're going to use a heat pump as your primary heater, and have a respectable amount of winter to deal with, it's worthwhile to pay a little extra money for a model with a high HSPF. On the other hand, if you haven't much winter to speak of, but a lot of air conditioning days, I'd pay more attention to the SEER side of the unit's rating.

Electric Furnaces and Combo Units

In addition to gas- and oil-fired furnaces and electric heat pumps, there are a couple of other choices for powering a forced air heating system.

One of these is an electric furnace, which isn't much more

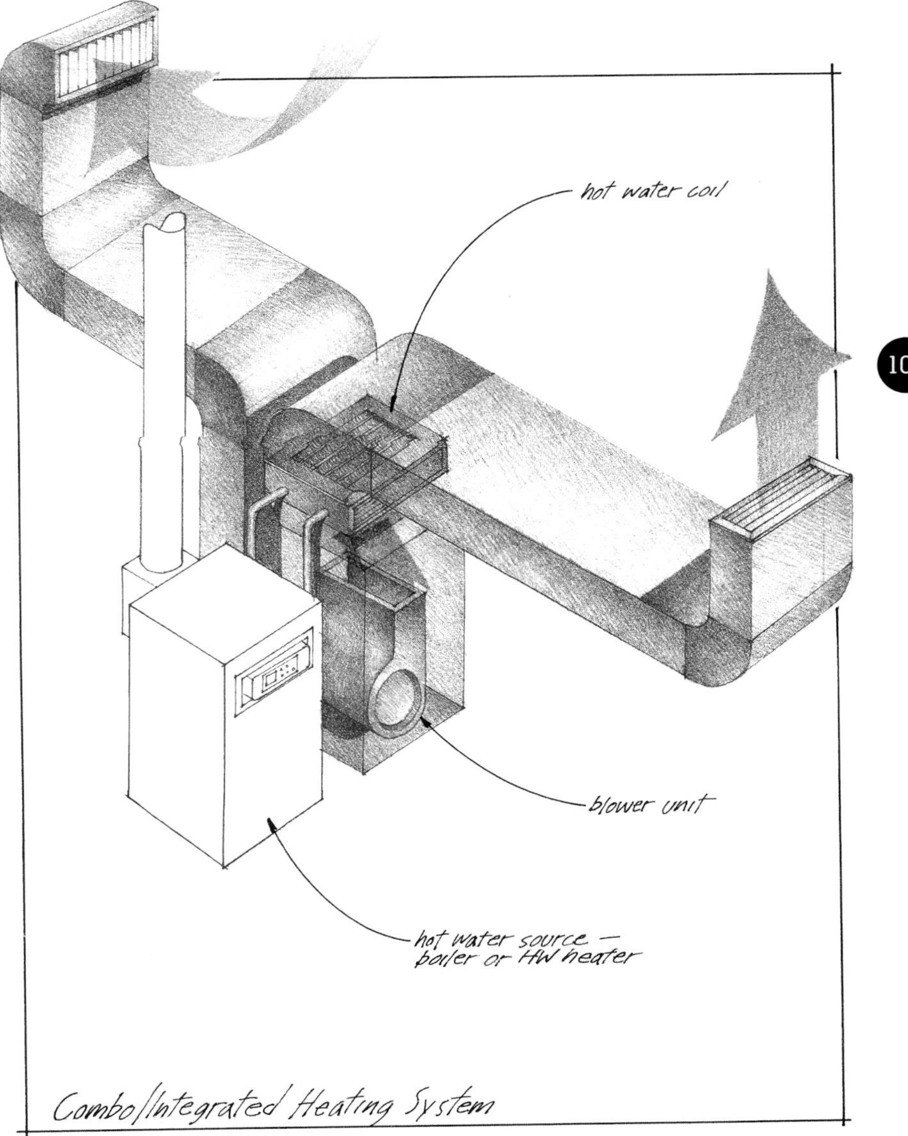

Combo/Integrated Heating System

than a bunch of electric resistance coils packed into an air handler. Because of the high cost of electricity in most places, it's hard for me to imagine a situation where I'd ever recommend an electric furnace — maybe, just maybe, in an area where the winter was short and sweet, and electricity prices cheap. Or in a house so tightly built and insulated that it wouldn't matter much how you heated it, since the bills are going to be insignificant anyway. It says a lot, I think, that most of my work with electric furnaces has been replacing them with something more economical.

Another alternative is to install a "combo system" — or "integrated system," as they're sometimes called — which is a kind of halfway creature between hydronic and forced air. Hydronic, you'll recall, refers to the use of hot water or steam in delivering heat.

At their heart, combo units have a high-efficiency water heater or boiler, which serves as the heat source for both hot water and space heating. As the diagram shows, hot water is pumped out of the water heater or boiler, through an insulated pipe, and into the air handler. Inside the air handler, the hot water winds its way through a serpentine heat exchanger, or coil. Air blowing over the hot coil is heated and distributed, via ducts and registers, through the home. Once the water leaves the air handler, it's piped back into the water heater or boiler for another circuit.

Using a water heater to provide both potable hot water and space heating is a bit of a new twist in the HVAC world. State Industries and Mor-Flo Industries, two of the companies that pioneered these so-called combo units, have had to wage a state-by-state fight to get code officials to accept the new technology. Forty-nine states (New York is the sole exception) have now revised their codes so that water heaters can be used for space heating.

Personally, I don't recommend using a water heater as the power source in integrated systems. I prefer to use a boiler in that role and design the system so that the boiler water and potable water can never mix.

The beauty and economy of this system is that instead of having two fuel lines, and two burners, and two flues, you get all of your space heating and hot water from one very efficient boiler.

We've used this type of integrated system on several "This Old House" projects, dating back to 1987. The most interesting was the Weatherbee Farm, where we ran *three* different loads off a single Burnham boiler. For space heating, we piped hot water from the boiler to two air handlers, one located in the basement, the other in the attic. The air handlers distributed warm air to the home's first and second floors, respectively, which gave us separate zones.

A second pipe from the boiler ran through a heat exchanger inside a superinsulated hot water storage tank — this replaced the conventional water heater and provided all of the family's hot water needs.

Finally, when we poured the new floor in the kitchen, we laid a serpentine loop of plastic pipe right into the slab. Hot water from the boiler, coursing through the in-floor pipe, provided radiant heat to the kitchen. I'll have more good things to say about combo systems and radiant underfloor heating in chapters to come.

Right Back Where We Started

In the end, of course, everything rests on the skill and integrity of your contractor. The new equipment I've described above is worthless if it isn't properly installed and serviced.

Trethewey's Tips: Top off That Tank

For those of you who heat with oil, I recommend that you have your storage tank filled in the spring, just after the heating season had ended. If the tank is left half full over the summer, water condenses inside it, causing the tank to rust. Water and rust can also get into the fuel line and cause problems with the burner on your furnace or boiler.

One way to find out if water has already condensed and pooled in the bottom of the tank is to put a smear of gray litmus paste on the dip stick and lower it to the bottom. If the paste turns purple, you've got water there. If the test indicates water more than about an inch deep, get a bucket and drain it out the bottom of the tank.

You may also save a few bucks by filling your oil tank in the spring, since the price per gallon tends to slip as the heating season winds down. And when that first cold snap comes through in the fall, you'll be all fueled up and ready to go.

- If you do install the unit yourself, here are some tips:
- Mount the thermostat on an interior wall about 5 feet off the floor.
- Don't use a wall that's facing unconditioned air space or that has pipes or flues in it.
- Avoid putting the unit near air outlets, radiators, or drafty spots.
- Don't place the thermostat in a closet or behind a door.
- Make sure the spot you choose doesn't get direct sunlight.

The contractor's vital role in all this was documented recently by Canadian researchers, working for Manitoba Energy & Mines, who discovered that many high-efficiency gas furnaces weren't delivering the comfort and energy savings their manufacturers promised. The reason: shoddy installation work.

The most common and grievous installation sin — dating back to the very dawn of the HVAC industry — is putting in equipment that has too much capacity for the job. The practice of oversizing equipment may have gotten its ill-fated start before the Civil War, when people believed in leaving a window open — even in the dead of winter — to let out "vitiated" air, which was the suspected cause of many diseases. To compensate for that always-open window, early-day contractors made it their practice to oversize heating systems by about 40 percent, and half the HVAC contractors in America are still doing it.

Why? Because the contractor figures that if the system is oversized by a healthy margin, he won't get any complaints from the customer. Furthermore, it's less work for him to guess at the size of the equipment, with a healthy "fudge factor" added in, than it is to do a proper load calculation. Also, bigger systems usually generate bigger profits, not just for the contractor, but for the manufacturers and wholesalers too.

If I'm making this sound like an international conspiracy, it's because that's the way it sometimes feels. A typical heat load calculation done here in Boston — and we do them all the time — might find a house needing 32 to 40 BTUs per square foot to heat it properly. But down in the basement — nine times out of ten — we find a furnace or boiler that's sized to deliver up to *twice* that much.

The results of this gross oversizing are several, and all of them are bad. First of all, oversized equipment costs you more up front. Second, instead of running efficiently for long periods, oversized equipment cycles on and off all the time, gulping extra fuel. (Imagine a high-performance car condemned to spend its life in stop-and-go traffic.) Third, oversized equipment can cause serious comfort problems and tends to be noisy. Last, oversized equipment always requires more maintenance and usually dies young.

There's no great mystery to sizing a heating system properly. All things considered, it's a lot easier than doing your income taxes. All a contractor needs to know is:

- The outside design temperature. This is the wintertime temperature extreme that the heating system will be designed to handle. For example, the outside design temperature for Anchorage, Alaska, is a frigid –18°F while Miami's is a balmy 48°F. A list of outside design temperatures, including dozens of U.S. cities, is readily available in HVAC handbooks and manuals.
- The heated volume of the house. This is the measure of the interior space of your home (above grade only) in cubic feet. All you need to calculate the number is a tape measure and some sixth-grade mathematics — width x length x height.
- The average wintertime infiltration rate. This is expressed in Air Changes per Hour (ACH) — that is, the number of times the home's total air volume turns over in an hour. For example, if the volume of a house is 12,000 cubic feet and the infiltration rate is 18,000 cubic feet per hour, the house would be rated at 1.5 ACH. Most contractors just guesstimate the ACH by inspecting the house and categorizing it — Poor, Average, or Best — on a standardized table. But the infiltration rate can be more accurately estimated by testing the house with a blower door.
- The surface areas and R-values of the home's floors, walls, ceilings, windows, and doors.
- A little information about the family's lifestyle.

Once your contractor has gathered these pieces of information, all he has to do is follow the sizing procedure that's in the handbook.

As a rule, it's better to precisely match the equipment to the load than to put in too much capacity. The one and only instance in which it might make sense to oversize HVAC equipment is if you're planning — with great certainty — to expand your house later on.

Of course, sizing the furnace isn't the only prerequisite to a comfortable and economical heating system. A good HVAC contractor has to understand how to install the supply and return ducts, with the registers properly sized and positioned, so that the right amount of conditioned air flows quietly into each room without creating drafts. To accomplish this, the contractor must also have an understanding of modern electronic controls.

●

Hydronic Heating Systems

Here in Boston — and over most of New England — we have a long tradition of heating our homes with hot water and steam. While warm air furnaces, heat pumps, and baseboard electric predominate in other parts of the country, hydronic heating is still the system of choice for most of us hard-headed Yankees.

One of the reasons we've done so many hot water heating systems on "This Old House" is for the very simple reason that most of the old houses we've worked on — being located in New England — came with boilers, pipes, and radiators already in place. To the extent that we could salvage what was already there, we were able to save time and money. Common sense will tell you that ripping out an entire hydronic system so that you can put in ductwork (or vice versa) is one of the most radical and expensive changes you can make to a house. On our budgets, it wasn't something we hankered to do.

But there are other reasons for the show's bias toward hot water heating, which stem in large part from my own experiences and preferences.

Most important, I'm convinced that hot water heating is more comfortable than forced air. By their very nature, forced air systems rely on *blowing* warm air into the house, which can make people feel uncomfortable and stir up dust. Hydronic systems, even those that use convecting radiators, tend to distribute heat more gently, with fewer drafts.

As I discussed in the last chapter, forced air systems tend to leak and operate out of balance, which pressurizes or depressurizes different parts of the house. This draws cold, dry air into the house, with brutal effects on comfort and economy. Hydronic systems can be out of balance too, of course, with comfort problems resulting. But there's no infiltration penalty related to depressurization.

I also like the extra precision and control that hot water heating allows. By its very nature, a hydronic system has to be free of leaks, so you start with a certain amount of confidence that the heat you get from the boiler will actually be delivered to the rooms where it's needed. (By contrast, forced air systems can lose a lot of heat between the furnace and the final destination.)

Moreover, the precision with which hot water systems can be controlled makes them relatively easy to zone, which is a big plus in comfort and economy. Thanks to the invention of thermostatically controlled radiator valves, even old-style radiators can be zoned without too much effort.

Hydronic systems can also conserve space inside the house. Since the heat-carrying capacity of water is 3,000 times greater than it is for air, hydronic systems can use small-diameter pipes to distribute heat, while forced air systems have to rely on large ducts. Many modern-day radiators are slim and flat, designed both to save space and look sharp. Hydronic heating systems that have underfloor distribution don't use any floor space at all.

Finally, hot water heating systems (but not steam) tend to be quieter than forced air because there's no blower turning on and off all the time and no air whistling through the ductwork.

HVAC contractors who make their livings installing forced air heating systems will be quick to argue that ductwork doesn't *have* to leak, that air registers, when correctly sized and positioned, don't *have* to produce drafts, that properly designed systems don't *have* to cause infiltration problems or make a lot of noise.

True enough. And when the industry starts installing systems like that — day in and day out — I'll be the first in line to retool my opinion.

However, let me be quick to add that hot water heating systems are by no means perfect. First of all, they typically cost more to install than forced air systems.

Second, hot water systems tend to be slower to respond to a change in the thermostat than forced air.

Third, hydronic systems can't be easily adapted to handle air conditioning, ventilation, or air filtering.

Finally, if the pipes get frozen or there's a leak in the system, you can end up with serious damage. (Leaks in forced air systems don't reveal themselves except through comfort problems, noise, and/or higher fuel bills.)

In today's America, 6.7 million single-family homes — about 10 percent — are heated hydronically. These are located mainly in the Northeast and across the northern tier of the country, especially in the cities.

How They Work

Hydronic systems are characterized by a central boiler or water heater, which uses gas, oil, electricity, propane, wood, or coal to heat water. The resulting hot water or steam is then distributed through a pipe to some sort of radiator, which releases heat into the

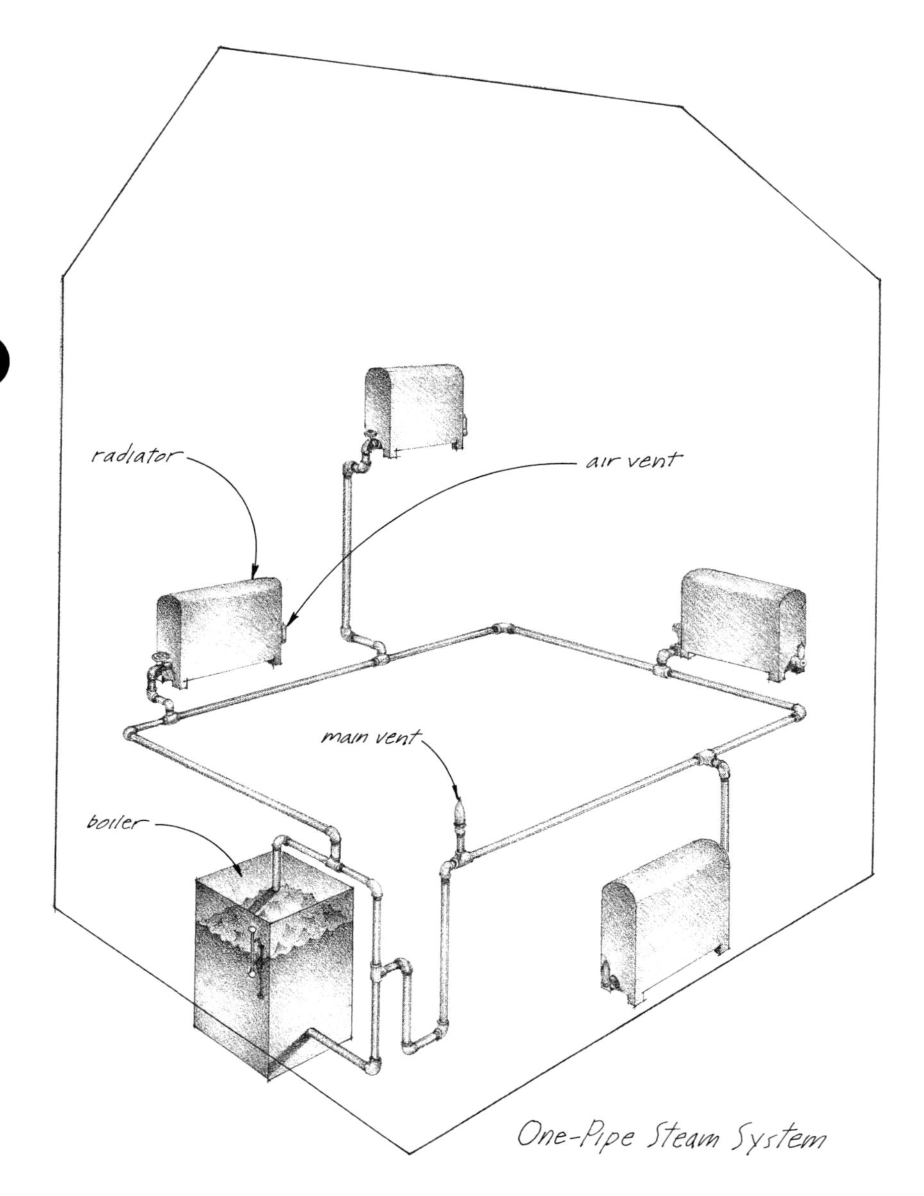

One-Pipe Steam System

room. The heating cycle is completed when the water or steam has given up its heat in the radiator and returns to the boiler.

Up until the mid-1930s, steam boilers were the popular choice for residential heating, employing the same technology — on a pint-size basis — that powered America's locomotives and factories.

The best of those early-day steam systems demonstrated a remarkable understanding of physics and very clever engineering. They worked well without pumps and sophisticated controls, just as long as there was someone around to periodically shovel in the coal and keep water in the boiler.

As shown in the drawing on page 000, steam generated in the boiler would rise through the pipes under its own pressure and

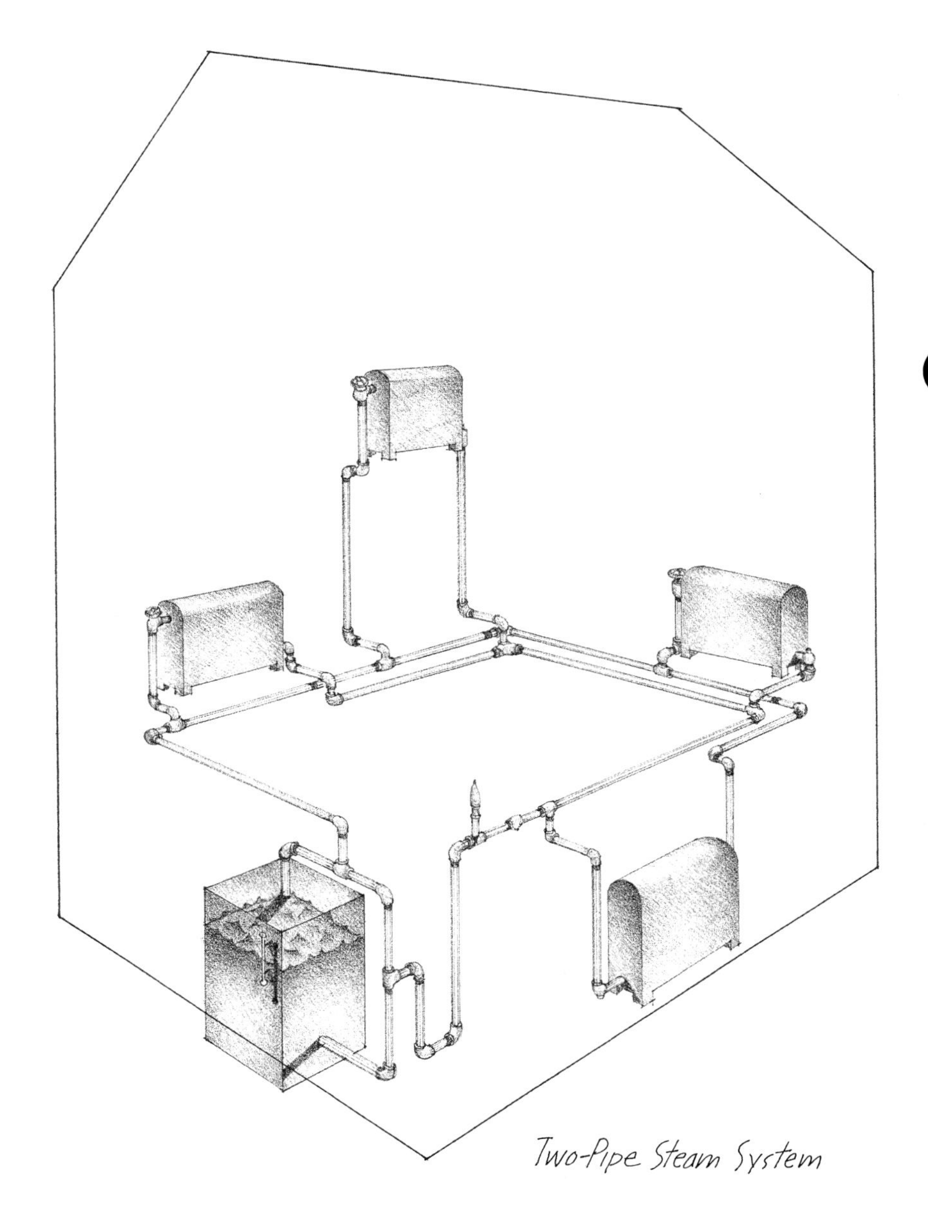

Two-Pipe Steam System

surge into the home's radiators. An air vent on each radiator remained open, which allowed the steam to enter, until the temperature rose high enough to close the vent automatically. As the steam inside the radiator cooled and condensed, the water collecting inside would either trickle back down the same pipe through which the steam had come (a one-pipe system) or drain down through a second pipe (a two-pipe system). In either case, the pipes had to be gently pitched so that gravity could return the water to the boiler.

Though the old steam systems are elegant in their simplicity, they have fundamental weaknesses, which have made them obsolete now for new residential applications. For one thing, a steam boiler has to raise the temperature of the water to 212°F before the system can even start to deliver heat. In other words, it runs

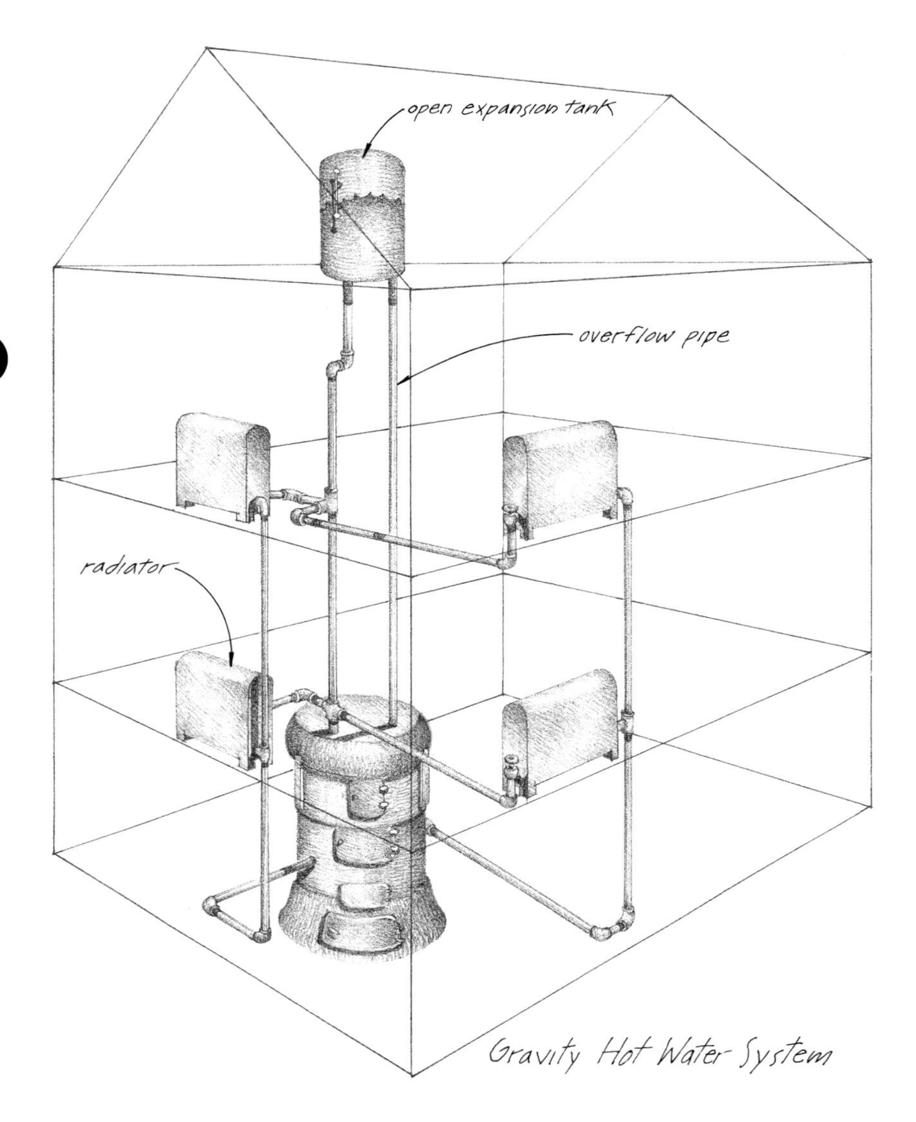

Gravity Hot Water System

HOT. Second, steam systems provide such quick and mammoth volumes of heat to the house, they can produce wild swings in indoor temperature. Last, steam systems can be noisy if they haven't been properly installed and maintained. Anyone who's ever lain awake at night listening to steam pipes bang and air valves hiss knows exactly what I'm talking about.

Gravity hot water systems are similar to steam, both in the way they look and the principles on which they operate. As with steam, hot water rises from the boiler into the home's pipes and radiators because that's what the laws of physics demand. As the hot water surrenders its heat to the radiator, it cools and falls, drawn back down into the boiler by gravity.

But there are some notable differences between steam and hot water. Whereas steam boilers need to be halfway full of air to work, hot water systems are completely filled with water, except for the air that's left in the expansion tank. (See page 121.) Second, since hot water systems don't have to boil to work, they can begin to deliver heat to the radiators at lower temperatures. A third difference, of course, is that hot water systems can be mechanically pumped, whereas steam relies exclusively on the natural forces of physics.

Over the years, the "This Old House" crew has tackled lots of old steam and gravity hot water systems, rehabilitating some of them, sentencing others to the junkyard. From a historical and slightly sentimental point of view, it breaks my heart to change some of these old gravity systems. Some of them are real works of art, installed by the grand old masters of the trade. They existed in a simpler time — or so I like to imagine — when folks spent more time at home and didn't mind the occasional trip to the basement to stoke up the boiler. The coal would burn slowly and heat would rise gently into the pipes and radiators.

But the old gravity systems, like a lot of other things in America, fell victim to our love of convenience. People discovered that by putting in a fuel oil tank or gas line, and equipping the boiler with a new burner, they could make their heating systems "automatic." No more stoking. No more dirty coal piled in the basement. No more worrying about pipes freezing.

The problem, of course, was that comfort often suffered in the process. The large pipes and radiators used in gravity systems were designed to handle the nice-and-easy heat that slow-burning coal produced. When homeowners switched to the more powerful fuels of oil or natural gas, they sometimes discovered that their nice-and-easy heating system had become a runaway freight train.

In my view, a lot of beautiful old gravity systems were massacred when they were converted to oil and gas. If your system was one of these victims of "modernization," don't give up; there are contractors out there who know enough and care enough to fix them right.

Apart from the introduction of oil and gas boilers that fired automatically, the most important change in hydronic heating has been the addition of pumps. Instead of relying on the laws of physics and the sure hand of gravity, modern systems use an electric circulator to push hot water out of the boiler, through the radiators, and back into the boiler again. Hence the name: *forced* hot water system.

The use of a circulating pump enables a hot water system to respond more quickly to the thermostat. It also makes it possible to use smaller pipes and radiators. (The old gravity systems employed large pipes and radiators so that water or steam moving through the

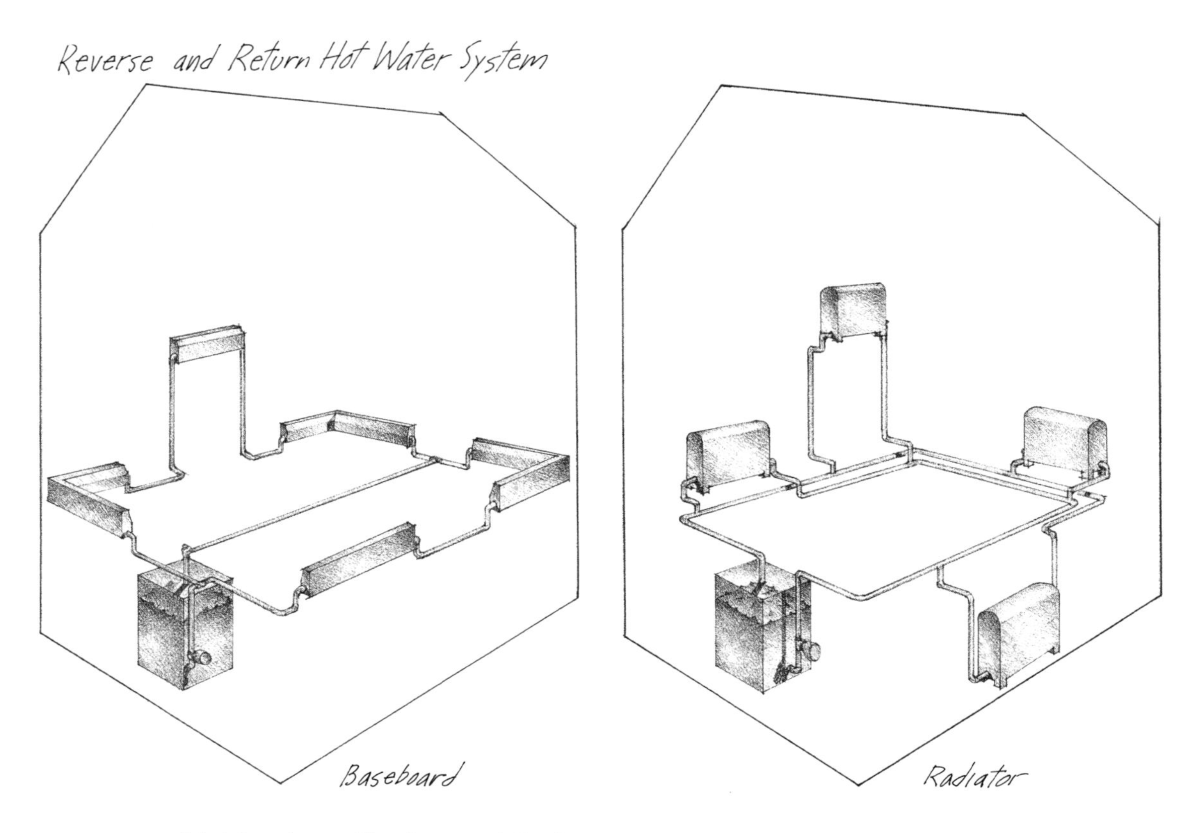

system wouldn't be slowed by friction.) And since forced hot water systems don't rely on gravity to return water to the boiler, you can locate the boiler wherever you want — even up in the attic.

Caring for What You've Got

I remember the first time we went down into the basement at Tug and Beth Yourgrau's old house in Melrose, Massachusetts. There, looking like a giant pancake, looming in the shadows, was the dirtiest boiler the world has ever known. I could tell by its design that it had originally been a coal burner, but had long since been converted to oil.

When I finally worked up nerve enough to open the combustion chamber and look in, it was so thickly covered with soot that its efficiency must have been down to 30 percent or so. In other words, for every dollar spent on fuel oil, 70 cents was being wasted.

If that ugly old beast-of-a-boiler wouldn't give a contractor nightmares, I don't know what would.

In the end, we decided to clean and tune the boiler thoroughly and leave it right where it was.

Why didn't we replace it?

Because it takes money to save money. In the best of all possible worlds, we would have yanked that boiler right out of there and put in a new high-efficiency model. Over time, the change

What the Doctor Ordered

Recommended Maintenance for Hydronic Systems	Hot Water Systems	Steam Systems	Can I Do It Myself?
Clean (or replace) fuel nozzle; clean the burner and combustion chamber; tune the system[1]	X	X	No
Run a combustion test to check boiler's efficiency and preclude any danger of backdraft	X	X	No
Check for leaks in the gas line	X[2]	X[2]	No
Check the thermostat	X	X	Yes[3]
Insulate the pipes	X	X[4]	Yes
Keep the radiators clean and free of obstructions	X	X	Yes
Change oil filter	X[5]		No
Manually adjust aquastat	X		Yes
Bleed air out of the system	X[6]		Yes
Lubricate the motor and pump[7]	X		Yes
Drain the expansion tank[8]	X		Yes
Drain sediment from boiler	X		Yes
Clean or replace clogged radiator vents[9]			Yes
Replace a bad steam trap[10]			No
Make sure the system has the proper amount of water			Yes

1. Once a year for oil-fired systems. Once every other year for gas.
2. Gas-fired systems only.
3. With an accurate thermometer and some patience you can detect a malfunctioning thermostat. But let your service rep clean or recalibrate it. If the thermostat needs replacement, you may be able to do it yourself. (See page 87.)
4. Compressed fiberglass only.
5. Oil-fired systems only.
6. Once or twice each heating season recommended.
7. Use oil sparingly, as recommended in the owner's manual. Newer circulators with sealed bearings don't require oil.
8. Older steel tanks only. Modern diaphragm-style tanks don't need maintenance.
9. One-pipe steam system.
10. Two-pipe steam system.

would have easily paid for itself in comfort and economy.

But since we didn't have the money in the budget to do that, we settled for the next best thing. By cleaning and tuning the boiler we raised its efficiency from 30 to 60 percent — no great shakes, to be sure, but a definite improvement and the best that we could afford under the circumstances.

I recommend that oil-fired boilers be professionally cleaned and tuned once a year. Gas-fired equipment, which generally burns cleaner, can be scheduled every other year.

To drive home this point about proper maintenance, consider the following: Just one millimeter of soot deposited inside the combustion chamber — that's .0394 of an inch — can drop the efficiency of an oil boiler by 6 percent.

If your "service" technician thinks chucking a soot stick into the combustion chamber is all there is to cleaning a boiler, you might want to look for someone new. The combustion chamber needs to be thoroughly brushed and vacuumed to do the job right.

Likewise, the old method of eyeballing the size and color of the flame isn't good enough anymore — a good service person will use a test instrument to adjust the flame properly.

The maintenance list on page 115 includes some other things that should be part of a professional service call.

Investing Sweat

If you'd like to tackle some of the simpler maintenance work yourself, here are some doable projects for hydronic heating systems.

Adjust the boiler temperature

Hot water heating systems have a control device called an "aquastat," which is usually found inside a metal box connected to the boiler. The aquastat tells the boiler, within an adjustable range, how hot to heat the water. During cold weather, an aquastat might be comfortably and economically set at 180°F (for the high limit) and 160°F (for the low).

But during periods of relatively mild winter weather, especially in the autumn and spring, there's no need for water that hot. By resetting the aquastat to 140°F on the high limit and 120°F on the low, you could save 5 to 10 percent on your fuel bills.

Later on I'll talk about the advantages of having an outdoor temperature sensor on your heating system so that those kinds of money-saving setbacks are made automatically.

But first a word of caution: It may not be desirable for you to lower the aquastat settings if you have a coil inside the boiler that's being used to heat your potable water. Cutting back the aquastat, in that case, could cause a shortage of hot water. This is only one of many disadvantages related to tankless coils, which will be discussed further in chapter 9.

A second word of caution: When you have the heating system serviced, make sure the technician doesn't leave the aquastat settings higher than you want. Some service people bump the settings way up as a matter of habit so that they won't get any callbacks.

Flush out the boiler

Some hydronic systems are built so tightly that they continue to

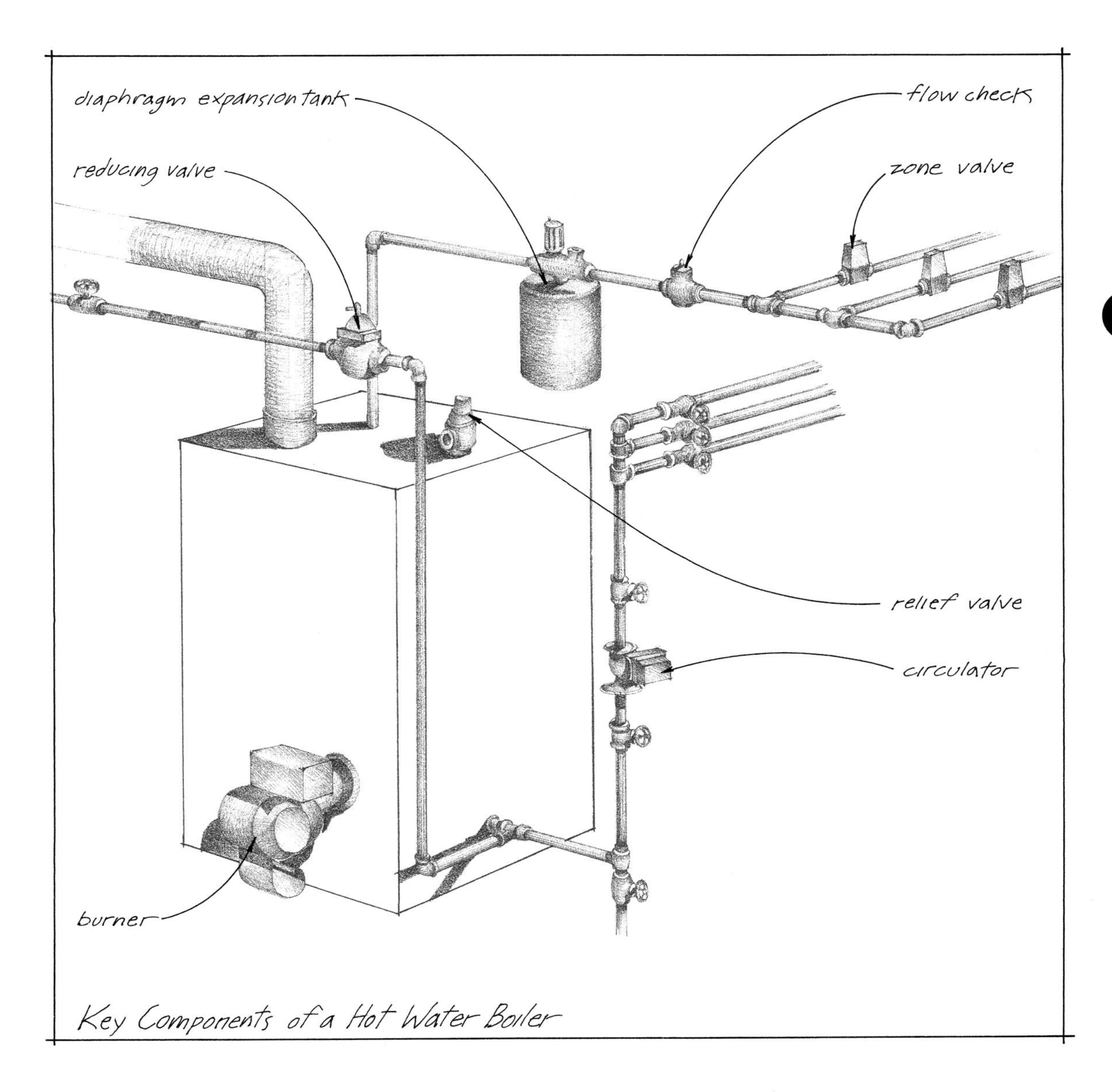

Key Components of a Hot Water Boiler

cycle year after year without losing any water from the system or collecting much sludge in the boiler. Others need to have new "make-up" water added frequently and accumulate lots of sludge.

As part of your annual maintenance routine, I'd recommend drawing a bucket or two of water out of the boiler to remove any sludge that's deposited there.

If you have a hot water heating system, start by turning off the burner and water supply to the boiler, and letting it cool for a few hours so that you won't scald yourself. Now position a bucket under the drain valve or drain cock on the boiler and draw off a bucket of water. (If there is more than one drain valve near the bottom of the boiler, take a bucket from each.) When you're done, open the water-supply valve and turn the boiler back on.

If there's a lot of sludge apparent, you should probably go ahead and flush the whole system. Start as you did before, with the water supply off and the boiler shut down and cooled. But instead of using a bucket, connect a hose to the drain valve and run it to an appropriate indoor or outdoor drain.

To get the system to drain properly, you'll first need to open the air vents on those radiators that are at the highest elevation in your house. Then open the valve or drain cock on the boiler and let the whole system flush.

Once all the water and sludge have drained out of the system, close the radiator valves and open the fresh-water-supply valve. Let some water flush through the boiler until it runs clean. Then close the drain valve or cock so that the system refills. Any air that's trapped in the system can be released later.

Even if it's summertime, I recommend that you turn the fuel back on and fire up the boiler for a while. Heating up the system will drive air (oxygen) out of the fresh water and help prevent rust.

To flush steam systems, start by making sure that there's water showing in the sight glass located on the front of the boiler. If there's none showing, it means that you need to add water.

Some boilers have an automatic fill control; others need to be filled manually. Obviously, if your system is supposed to fill automatically, yet isn't full, something is wrong. Leave the system off and call a repair person.

If your system is the type that fills manually, open the fill valve that's located on the pipe running between the water meter and the boiler. Leave it open until the sight glass is half full.

Always let the boiler cool down for a few hours before you add water. I've seen boilers cracked — that is, ruined — because people pumped cold water into them while they were still hot.

If the sight glass *is* showing water, you're all set to flush out the boiler. First, turn up the thermostat so that the boiler comes on. Once it has, hold a bucket underneath the drain valve that's located near the bottom of the boiler and open the valve. (If your system has more than one drain valve on the boiler, draw some water from each one.)

As water and sludge drain out, the boiler should automatically shut itself off. The automatic shutoff is a built-in safety switch designed to protect the boiler (and you) against low-water conditions. If the system doesn't automatically turn itself off, and the sight glass is showing empty, turn the system off at the thermostat and call a repair person.

Once the old water and sludge have drained out of the system, let some fresh water flush through it. Then close the drain

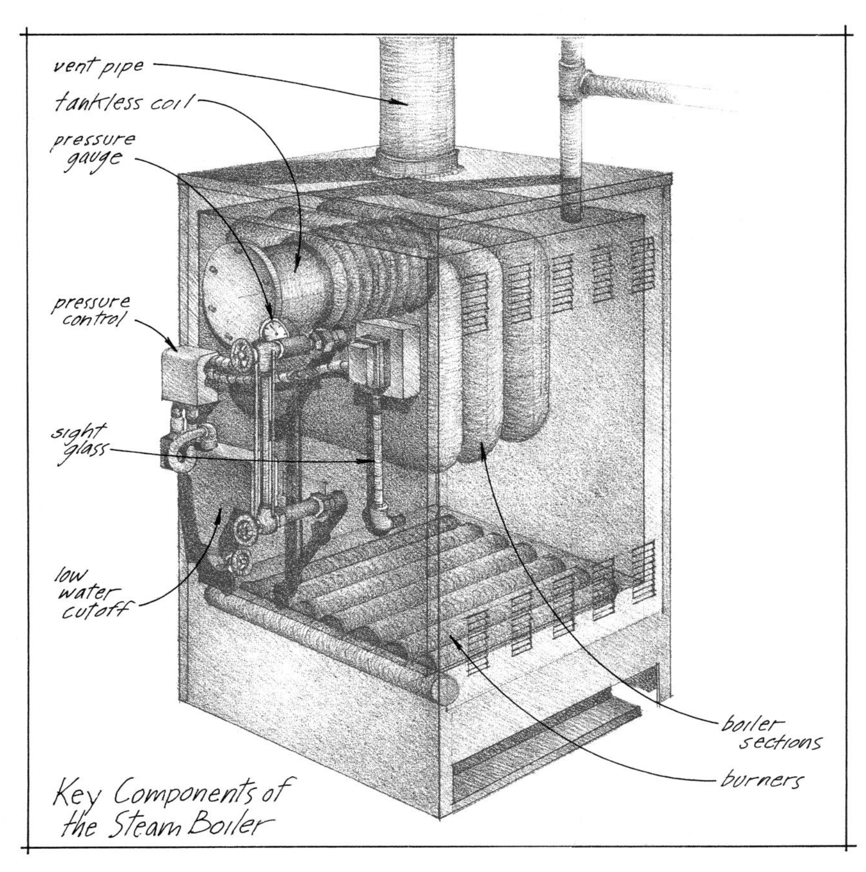

Key Components of the Steam Boiler

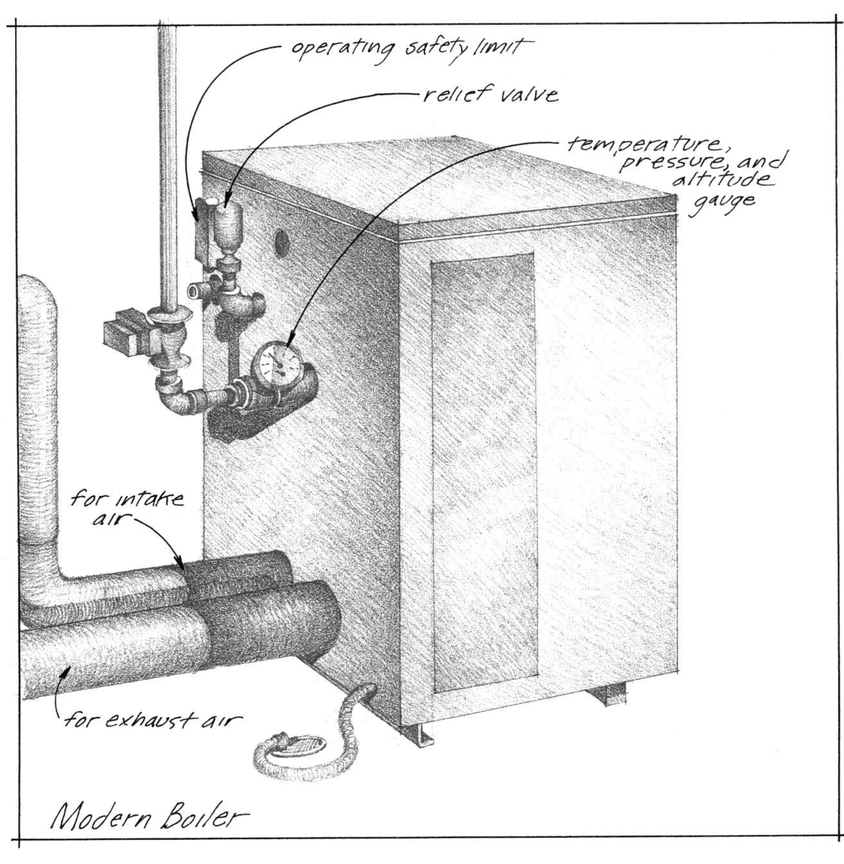

Modern Boiler

valve. If your system has an automatic refill control, it should automatically refill the boiler to its proper level. If you're using a mechanical feed valve, you'll need to leave it open until the sight glass is half full.

As a matter of routine, it's a good idea to check the water level in a steam boiler about once a week. And it's worth repeating: Be sure to let the boiler cool before you add any water.

For steam and hot water boilers that are really sludged up, you should probably hire a pro to "power flush" the system. By using special chemicals and super hot water, even hard-to-remove sludge can be flushed out.

Drain the expansion tank

Because water wants to expand when it's heated, hot water heating systems are fitted with a special tank so that the water can expand in a safe and controlled fashion.

Early-day expansion tanks were located somewhere above the highest radiator, usually in the attic. The top of the tank was open, allowing air to move in and out of the tank as the water level and temperature in the system changed. They worked beautifully. The next generation of expansion tanks were plain steel cylinders — 3 to 5 feet long — that were typically hung above the boiler. These tanks were designed to operate under pressure, half full of water, half of air. The problem with this closed tank arrangement is that the pressure forces air into the water. And when that water gets up into the radiators, where pressure and temperature drop, it burps air into the radiators. As a radiator fills with air, it loses efficiency and can start to make noise.

Since necessity truly is the mother of invention, it wasn't long before inventors came up with a better tank. Modern-day expansion tanks have a neoprene bladder or diaphragm — impervious to air — that expands or compresses with the changing volume of the water. So water and air need never mix.

While most of the newer style expansion tanks don't need to be drained, most of the older models — distinguished by a drain valve on their bottom — should be flushed out once a year.

Draining the tank is simple:

First, isolate the tank from the rest of the system by closing the shutoff valve. Second, attach a hose to the drain on the bottom of the tank. (Or position a bucket underneath it.) Third, open the valve. (Some models won't drain properly unless you also open a small vacuum-breaker plug that's built into the tank.)

After the water and sludge have drained out, simply reverse the process.

Incidentally, replacing an old-style expansion tank with one

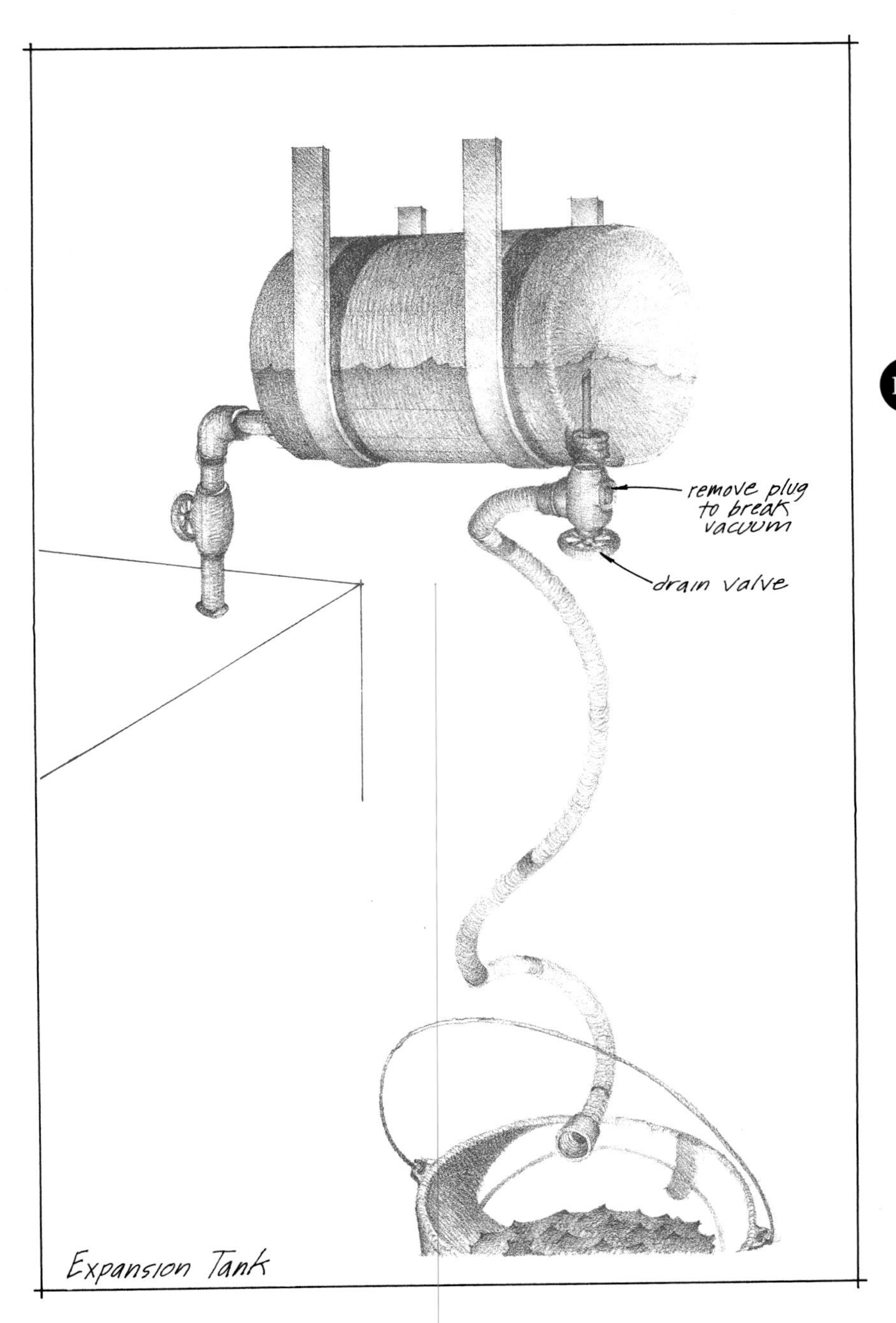

of the new diaphragm types is usually a worthwhile and not-too-expensive retrofit.

Bleed the radiators

When air gets trapped inside a radiator and prevents it from filling completely with hot water, you end up with a cold spot in your house. You can resolve this by bleeding the air out of each radiator once or twice each heating season.

It's a simple task. Hold a pan under the valve and open it with a radiator key. (You can buy a key at your local hardware store for about 25 cents.) When all the air has escaped from the radiator and only water comes out, shut the valve.

If you've already performed this tedious little chore too many times, you might want to invest in a nifty little gadget called a Spirovent. As I demonstrated on the show a few years back, Spirovent is installed as a permanent fixture in the piping near the boiler. Water passing through it is forced through a tightly wound spiral of wires that forces air bubbles out of solution and vents them to the room. Presto! No more need to bleed.

Spirovent sells for about $100, with installation running another $50 or $75. But since time really *is* money, the unit will eventually pay for itself in minutes saved.

There's no need to bleed the radiators on a steam system, of course, because they're *supposed* to have air in them. But steam radiators do have air vents — small metal cylinders built into the side of each radiator — that control the flow of steam. These vents have a nasty habit of clogging over time, which makes them hiss and drip and spit steam.

When that happens, try removing the vent (with the steam turned *off!*) and soaking it in vinegar or boiling it in water. If that doesn't unclog the vent, spring for a new one, which will run you $10 to $15.

Oil the circulator

At the start of each heating season take a few minutes to oil the motor and pump assembly on your system. This is done by placing a couple drops of lightweight oil into the oil cups that are built into the motor and pump housing; be sure not to overlubricate them. Some of the newer equipment doesn't have oil cups, because the bearings are sealed and self-lubricating.

Insulate the pipes

To insure that the heat generated in the boiler gets to its final destination, the distribution pipes should be carefully insulated. If you have a hot water system, wrap the pipes with specially formed tubes made out of foam or compressed fiberglass. These wraparound lengths of tubing cost 30 to 80 cents a foot, but will save you about 50 cents per foot a year in energy. Pipe insulation will also reduce or eliminate "water hammer," that loud, clanging noise that cold pipes can make when hot water or steam surges suddenly through them.

Clean the pipes off before you insulate them. If the insulation tubes aren't self-adhering, you can tape them along the seams and around the joints. The insulation should not touch or interfere with the operation of pumps, valves, pressure-relief devices, vents, or other equipment.

I also recommend insulating the water pipes returning to the boiler so that they won't sweat and drip.

If you have a steam heating system, be sure to use compressed fiberglass insulation on the pipes rather than foam. Steam pipes can get hot enough to *melt* foam-style insulation.

Many old steam systems were insulated — boiler and pipes both — with asbestos, which presents a deadly respiratory hazard if it's broken loose and dispersed in the air. If your steam pipes or boiler are covered with a white clothlike sheathing, leave it right where it is. If pieces are broken or missing, call a company that specializes in handling asbestos.

This whole topic of asbestos is laced with irony and dark humor as far as I'm concerned. Twenty years ago, when I was an apprentice, I used to pour big sacks of asbestos into a vat and prepare it with water for application. "Mix that asbestos up really well, Richard," the foreman used to tell me. And I did. Usually with no gloves or mask.

Now, on some of our jobs, I find myself working with asbestos-removal companies that treat this stuff like radioactive waste. On the old house out in Wayland, Massachusetts, which belongs to Chris and Joan Hagger, a crew of specialists was called in to remove *three little pieces* of asbestos. The retail value of that work was $1,300!

I only wish I'd been paid at that rate when I was *installing* the stuff. Boy, how the world can change on you.

Radiators

As you can see from the assortment of styles we've presented on page 124, radiators come in just about every size, shape, and color imaginable.

Some of the old cast-iron radiators, with their ornate moldings, are genuine works of art. And if they're properly maintained, they work just fine. Yet I also admire the sleek lines, bright colors, and radiant comfort of the new European-style radiators, which are made out of aluminum, copper, and high-temperature plastics.

Most radiators use a combination of convection and radiation to heat a room. Convection, as you'll recall from chapter 2, is the transfer of heat from a solid to a gas, in this case from a hot metal surface to moving air. Radiation is the transfer of heat by electromagnetic waves, which move from a warm object to a cooler one without noticeably warming the air in between.

The radiator's design will determine how much heat is released through convection and how much is given off through radiation.

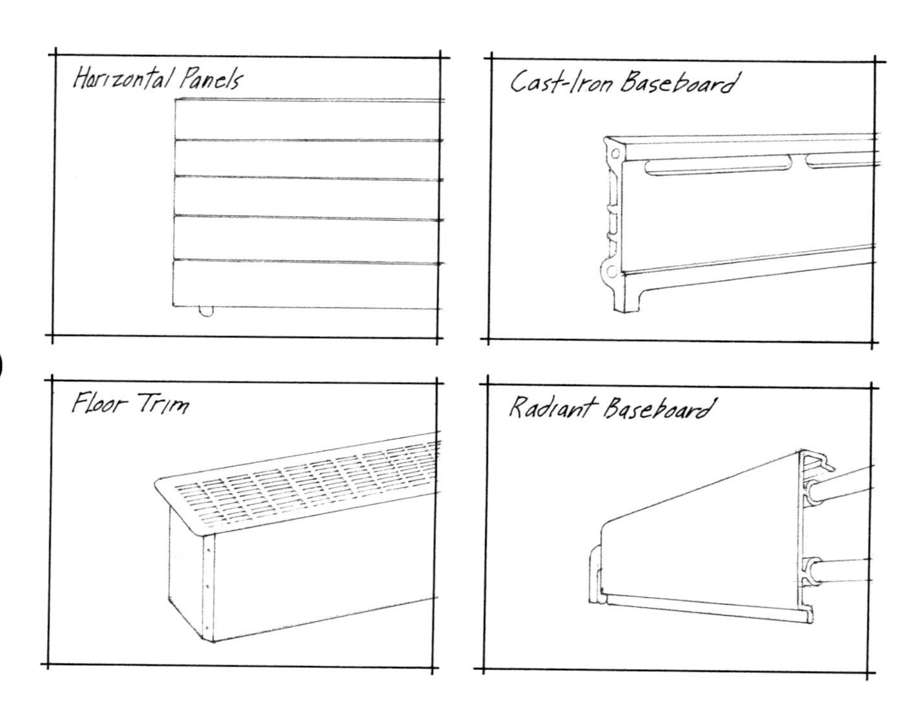

For example, baseboard radiators that have a copper pipe running down their length ringed with fins (hence the name "finned-tube heating") are designed to heat the room mostly by convection. The fins serve to increase the surface area of the radiator and promote air movement.

Low-temperature panel radiators, on the other hand, have a plastic or metal water tube encased in steel or aluminum and are designed to deliver most of their heat by radiation.

Copper-finned baseboard radiators, which have been around since the 1940s, are the relatively inexpensive, dependable Chevrolets of the industry. But they take up a fair amount of floor space around the perimeter of the room and, while not exactly ugly, aren't recognized for their beauty either.

By contrast, the flat panel radiators that originated in Europe might be described as the expensive, dependable Volvos of the industry. Some styles are so sleek and inconspicuous — designed, for example, to look like a pine baseboard — that the casual eye doesn't even realize they're radiators. Others are colored bright and sassy, in the contemporary style, to make a conspicuous and beautiful design statement.

In Lynn and Barbara Wickwire's home, in Concord, Massachusetts, we heated the upstairs bathroom with a stylish flat panel radiator that doubled as a towel warmer. Imagine stepping out of the shower into a toasty bathroom with a preheated towel at your fingertips.

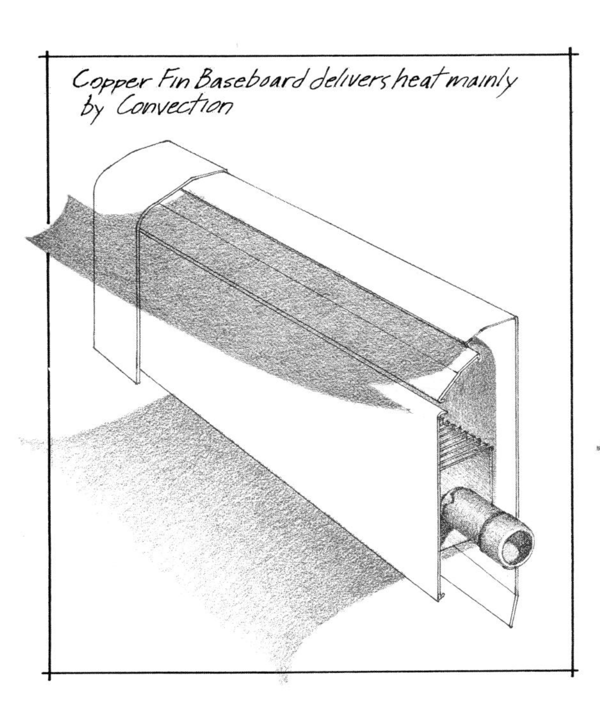

Copper Fin Baseboard delivers heat mainly by Convection

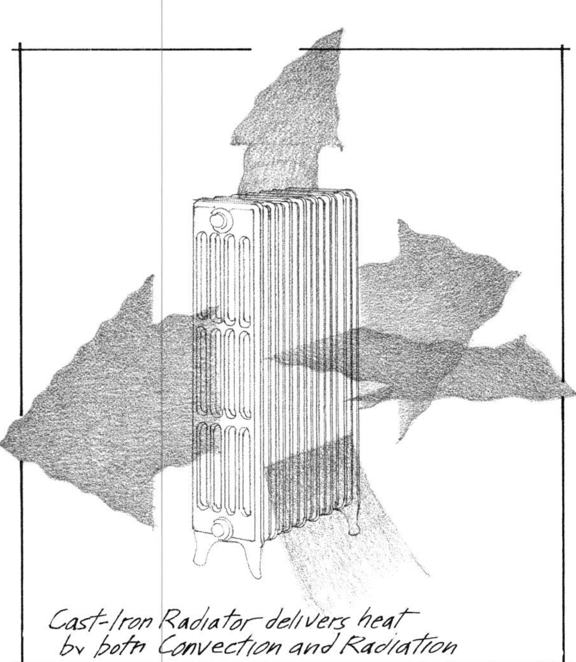

Cast-Iron Radiator delivers heat by both Convection and Radiation

While the looks and purchase price of radiators are important to most people, they aren't everything. Your HVAC contractor must understand how different types of radiators perform, both in the amount and type of heat they deliver, so that your system is both comfortable and economical to operate.

There's not much to add here about radiator maintenance. I've already explained the importance of releasing the air out of hot water systems and keeping the air vents on steam system radiators unclogged. Apart from that, be sure to keep all radiators clean and free of obstructions. Baseboard radiators, in particular, mustn't be blocked by close-in furniture or carpeting, since they need to pull air up across the radiator in order to be effective.

If you paint your radiators, be sure to use paint that will hold up under the heat. Though metallic silver paint may look beautiful — especially on those old ornamental upright models — the best color as far as heating efficiency is concerned is dull black.

Radiant Floors

Anyone who's watched "This Old House" for very long knows that I'm a big fan of in-floor hydronic heating. Rather than having unit or baseboard radiators positioned around the room, in-floor heating relies on a serpentine loop of hot water lines that are either embedded in the cement slab when it's poured or looped through the wooden joists that are under the floor. What this does, in effect, is turn the whole floor into a radiator.

Top: In our Lexington, Massachusetts, house, mechanic Tony Lucas (left) lays down the new subfloor while I insert cross-linked poly tubing into the precut grooves. Note in the detail below how the groove is caulked with a bead of silicone before the tubing is pressed flush into the new subfloor.

Bottom: Videotaping in the great room, in Lexington, Massachusetts, "This Old House" host Steve Thomas and I share some installation and operating tips on radiant floor heating. The floor was subsequently finished with laminated wood.

When I first looked at this technology some years back, I was both encouraged and scared. On the encouraging side, I discovered that the idea of heating a room from the floor up goes all the way back to the Romans, who were building "hypocaust" heating systems for their public baths as early as A.D. 60.

Some of you may remember the show in which we toured the famous Roman baths in Bath, England, which had underfloor heating in certain rooms. As you can see from the pictures here, the design was quite ingenious. And the Romans' original pool — more than 1,900 years old now — is still watertight!

While the advantages of radiant floor heating have been apparent for centuries, and human beings have been dogged in trying to perfect the idea, there have been some serious glitches along the way.

What scared me most was the dismal experience we had with these in-floor systems back in the 1940s and '50s, when thousands of them were installed in the United States.

It was common practice in those days to use a continuous loop of steel or copper pipe, which was embedded in the slab of a new home and connected to the boiler. Unfortunately, there were serious weaknesses in the design, installation, and — most especially — the materials used in those systems.

To begin with, the controls were primitive. Instead of keeping a continuous and even flow of warm water pushing through the slab, the controls would cycle the boiler on and off, sending a slug of hot water through the slab, then letting it cool.

To make matters worse, contractors ran the pipes in and out of the slab in such a way that it created stress points, which were destined to fail.

Last but not least, the lime in the cement began to corrode the copper and steel pipe — the more lime there was in the original mix, the faster the corrosion advanced.

Within a few short years, many of those early in-floor hydronic systems started to leak and had to be sealed off and replaced. The technology fell into rapid and seemingly permanent decline.

In the late 1960s a new technology — developed for use in telecommunications — gave in-floor hydronic heating a new lease on life. While looking for a new plastic tough enough to use as a sheathing for undersea telephone cable, researchers discovered cross-linked polyethylene.

Unique among plastics, the cross-linked poly can withstand wide and repeated swings in temperature without becoming brittle or mushy. The plastic also has a diffusion barrier that keeps oxygen

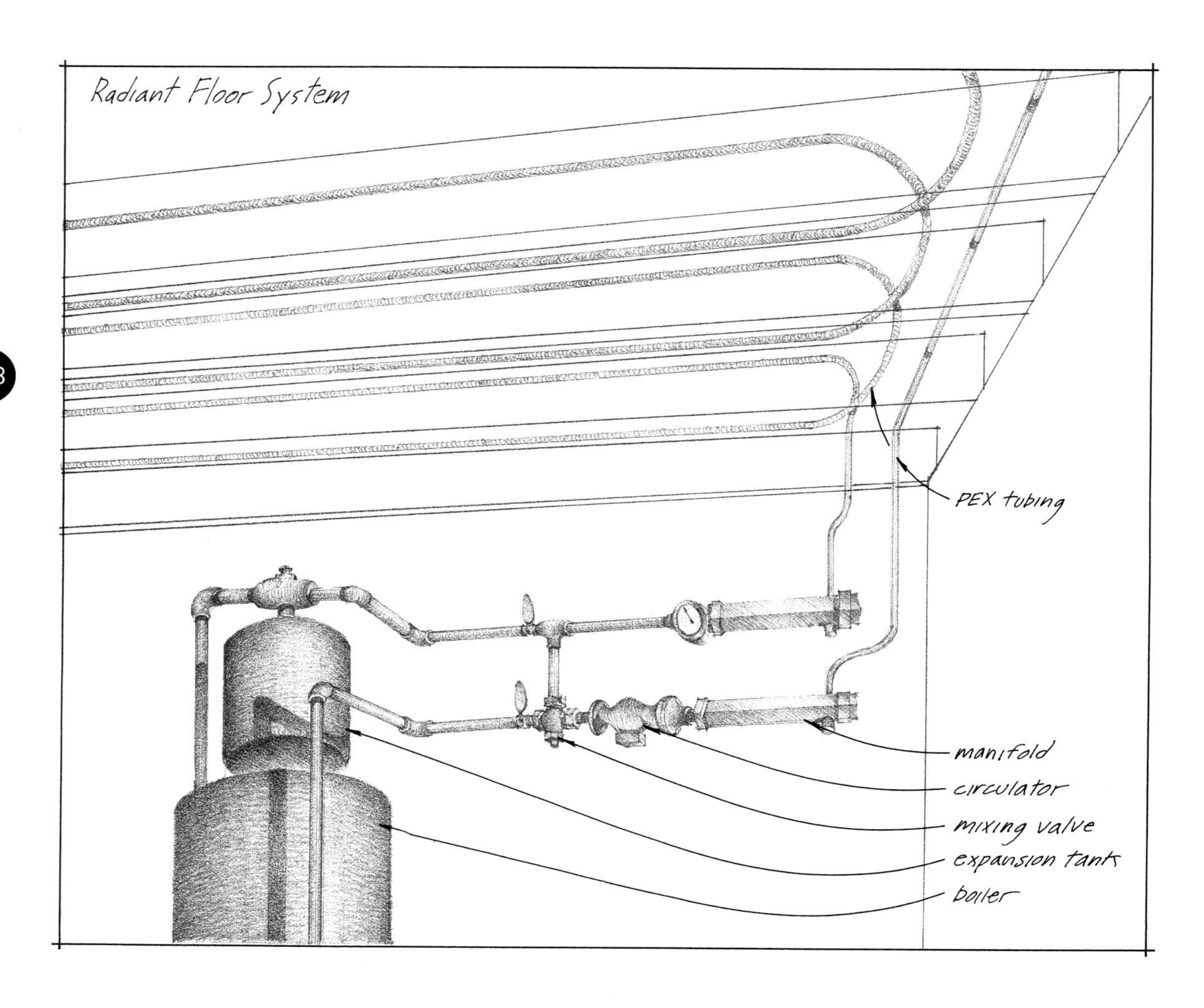

from passing through its walls. In practical terms, that means the pump, valves, and fittings in the heating system are protected from rust.

Now, everyone knows that plastic tubing isn't going to transfer heat into the floor as well as copper or steel pipe would. But in this case, it's durability that interests me most. Accelerated time tests, conducted by the independent laboratories that write standards for German industry, suggest that cross-linked polyethylene can last 200 years inside a slab. Real-life installations, in place without a failure since 1969, reinforce the test results.

In any event, I'm confident that the polyethylene tubing we installed in the slab of the Wickwires' home, in Concord, Massachusetts, and for James and Joe Anna Asher, in Santa Fe, New Mexico, will last as long as the houses themselves.

For those particular jobs, we used polyethylene tubing manufactured by Stadler, Inc., but you can get it from other companies too, including Wirsbo, Heatlink, Rehau, and EHT, Inc.

I would discourage homeowners and contractors from using other types of plastic or rubber tubing for in-floor systems. They've got a mixed track record for this type of application and may become brittle or otherwise fail after only 15 or 20 years. I'll repeat it loud and clear: If you're going to install in-floor hydronic heating, you have to do it with confidence — to me, that means cross-linked polyethylene with an oxygen diffusion barrier.

The other keys to successful in-floor heating are a properly sized boiler or hot water heater and a smart and reliable control system. The goal is to supply the slab (or subfloor) with a steady-but-gentle supply of hot water. Typical circulation temperatures run between 80 and 130°F, which is much cooler than other types of hydronic systems. I would strongly recommend that the controls be equipped with an outdoor temperature sensor, which helps the system anticipate the family's heating needs so that it can increase or decrease the water temperature accordingly.

Here are the three basic installation methods for in-floor heating.

- With new foundation work: Once the home's gravel drainage bed, vapor barrier, underslab insulation, and concrete reinforcing mesh are in place, the polyethylene tubing for the in-floor heating system is laid out. The concrete slab is then poured over the top of it, leveled, and cured.
- With new floors: The polyethylene tubing is laid out on top of the wooden subfloor and a slurry of lightweight concrete is poured over it. If the subfloor sits over unconditioned space, it should be insulated underneath. Another approach, which we used in my own kitchen, is to nail ¾-inch by 8-inch plywood strips or "sleepers" to the existing floor, spaced evenly apart, and then lay a new, elevated floor on top of them. The loops of polyethylene tubing run between the sleepers — that is, in the cavities between the old subfloor and the new, elevated floor. We used special metal sleeves to hold the tubing in place and reflect heat upward into the floor.
- Retrofit through existing floor joists: When it's impractical to pour a new concrete floor or use sleepers to raise the floor, the tubing can be installed on the underside of the floor, provided the joists are accessible from the basement or crawl space below. Holes are drilled through the joists so that the polyethylene tubing can be threaded through them. Aluminum sleeves fitted around the tubing hold it in place and direct heat upward into the floor. Insulation underneath prevents heat loss to the space below.

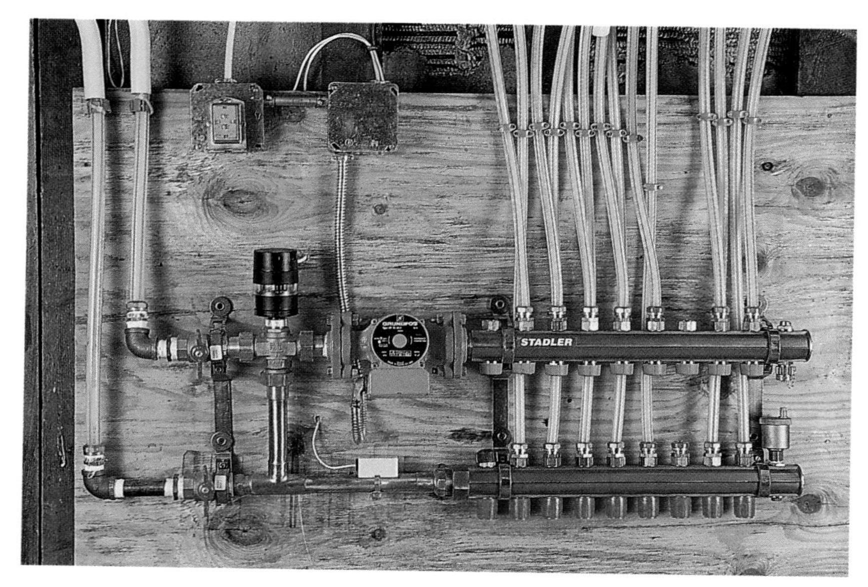

Top: It was fairly simple to add radiant floor heating to our project house in Wayland, Massachusetts, since we had easy access, through the basement, to the wooden floors above. Note the way the cross-linked poly tubing is looped through the floor joists and held in place by the aluminum reflector plates. The underside of the cavity was later filled with insulation.

Bottom: The new radiant floor heating system in Wayland, Massachusetts, was fitted with a control, pump, mixing valve, and manifold to distribute hot water to the proper zones.

With any of these methods, it's essential that the tubing be properly spaced. While these details are best left to a pro, you may be able to shave dollars off the job by doing some of the grunt work yourself.

One of the nicest things about in-floor hydronic heating is that it can be mixed and matched with other distribution methods. For example, you could install in-floor radiant heating in your new kitchen or family room without having to change the radiators in other parts of the house. I've listed some other advantages (and disadvantages as well) in the sidebar.

Currently, in-floor hydronic heating accounts for less than one half of one percent of the home heating systems in the United

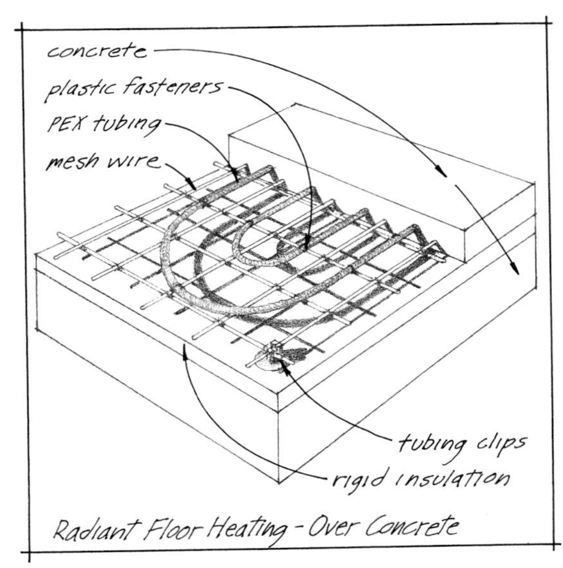

Radiant Floor Heating - Over Concrete

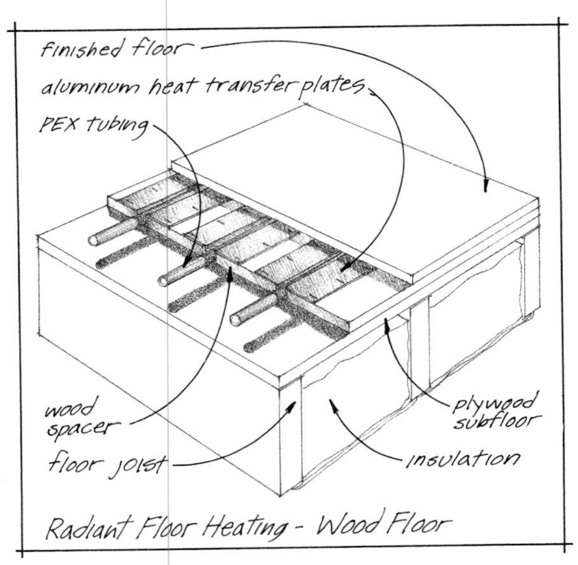

Radiant Floor Heating - Wood Floor

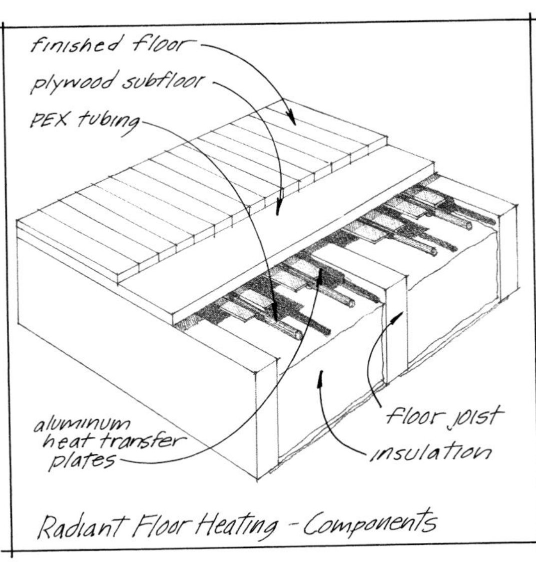

Radiant Floor Heating - Components

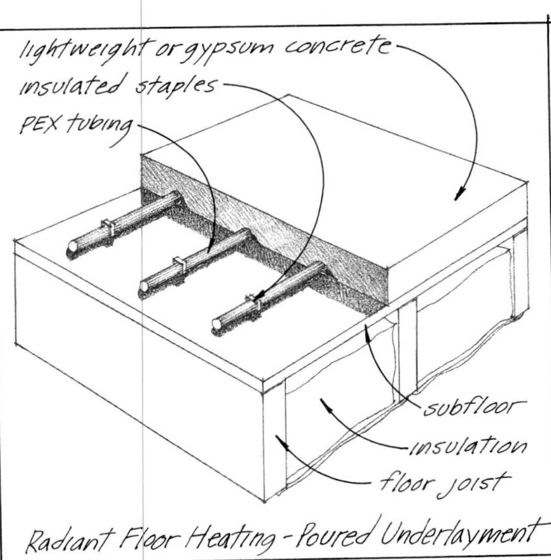

Radiant Floor Heating - Poured Underlayment

States. But I think its many advantages are going to propel it into the mainstream in the years ahead. In-floor heating is already the system of choice in Germany, where it accounts for 60 to 70 percent of all new systems installed.

Patchwork Solutions

Homeowners who are faced with serious comfort problems, high costs, or both, sometimes make rash decisions that they later regret. Before you put a "For Sale" sign out in front of your house or go into deep hock to buy a whole new heating system, I hope you'll consider some ways to make cost-effective improvements to your heating system. The chart on page 133 lists nine very promising options,

Pros & Cons: In-Floor Hydronic Heating

Pros

- Customers say (and I agree) that in-floor radiant heat is more comfortable than other heat distribution methods.
- People like being warmed "feet first" rather than having heat emit from the walls or ceiling. The floor feels warm, even on bare feet.
- There is less temperature stratification. In-floor heating counteracts cold air's natural tendency to settle on the floor.
- There are fewer hot and cold spots, and fewer drafts.
- The system is invisible. The floor and walls are uncluttered with unit or baseboard radiators or air registers. Furniture can go anywhere.
- The system is easily zoned, which enhances comfort and saves money.
- Because in-floor heating functions best with a steady flow of low-temperature water, the boiler can run in its most economical mode, with little cycling.
- It can deliver significant energy savings compared to a typical forced air system. (And marginal savings compared to other types of hydronic radiators.)
- It can operate with any fuel source that can heat water, including natural gas, oil, electricity, or solar.

Cons

- It's more expensive to install than most other system types.
- It may not be practical to retrofit in some cases.
- The pool of contractors who are experienced with in-floor radiant heating is still very small.
- The system responds slowly to the thermostat. (An outdoor temperature sensor helps address this shortcoming.)
- The system does not recommend itself to wall-to-wall carpeting, which has the negative effect of insulating the floor surface. (Imagine putting a sweater over your radiator!) If you must have carpet, avoid those with thick underlayments and shag or plush weaves.
- If pipes freeze or there is a leak, there can be serious damage.

some of which we've demonstrated on the show.

One of the most common problems with aging hydronic systems is that they frequently operate out of balance, producing hot and cold spots in different parts of the house. I can drive along Boylston Street here in Boston on a winter morning — with the temperature outside at 10° or 15°F — and see 300 open windows in the space of a mile.

Why? Because the rooms in some of those old brownstones are either frigid to the point where you can see your breath or so drastically overheated that the windows have to be left open to keep people from melting.

If your house has a hot water heating system, you may be able to resolve this kind of imbalance by adjusting the radiator valves. This is accomplished by cutting back the hot water to those radiators that are producing plenty of heat so that more hot water can be directed to the weak sisters, which are run with their valves wide open.

You can't make this kind of adjustment on a steam system, because the radiator valves that control incoming steam have to be left wide open to work properly. You can, however, change the size of a radiator's air vent. A replacement vent with a *larger* orifice would permit steam to enter the radiator more quickly, providing

Profitable Patches: For Boilers

Item[1]	Cost	Estimated Fuel Savings[2]	Comments
Install thermostatic radiator	$35–$75 per radiator	10–15%	Can be a do-it-yourself project on one-pipe steam system. Professional installation on two-pipe steam and hot water heating systems will cost about $200 per radiator.
Weather-responsive control (outdoor reset control)	$300–$1,000	5–25%	Hot water boilers only. Adjusts the boiler water temperature according to outdoor temperature. Savings are highest when used with a setback thermostat and thermostatic radiator valves.
Replace fuel nozzle with a smaller one	$0–$60[3]	2–10%	Easier and less expensive with oil boilers than with gas. Not applicable to steam heating systems.
Flame-retention head oil burner[4]	$250–$600	10–25%	Oil-fired boilers only. A better but more expensive option than downsizing the nozzle.
Replace pilot light with electric ignition	$150–$300	5–10%	Gas boilers only.
Install automatic vent damper	$250–$400	3–15%	Closes off the flue when boiler quits firing to reduce stand-by heat loss up the stack. Savings are usually greater with oil than with gas.
Install a clock thermostat	$40–$280	10–20%	Enables you to lower the temperature automatically while you're sleeping or away from home.[5]
Install gas power burner	$400–$600	10–20%	For converting old oil and coal systems to gas.
Radiator reflectors	$10–$100	Varies	A simple do-it-yourself project. By mounting galvanized sheet metal, aluminum foil, foil-faced insulation, or metallized film to the walls behind radiators (or convectors), heat loss through the walls is reduced.

1. Consult a good HVAC contractor or energy auditor about the cost-effectiveness of modifying your boiler versus buying a new one.
2. Savings on these items are *not* cumulative.
3. On oil-fired boilers (but not gas) the nozzle is sometimes downsized free as part of a tune-up.
4. May require downsizing the combustion chamber.
5. Do-it-yourself installation is possible on some heating and cooling systems. See page 87.

faster heat to the room, while a vent with a *smaller* orifice would slow the rate at which steam could enter.

A better solution by far — one I've demonstrated often on "This Old House" — is to equip each radiator with its own thermostatically controlled valve. What this does, in effect, is to turn every room that's equipped with a radiator into a separate temperature zone that can be individually controlled. As I explained in chapter 6, zoning can produce remarkable improvements in both comfort and economy.

Thermostatic radiator valves can be retrofitted to either steam or hot water systems, as long as the radiators aren't connected in series. An able do-it-yourselfer could zone a one-pipe

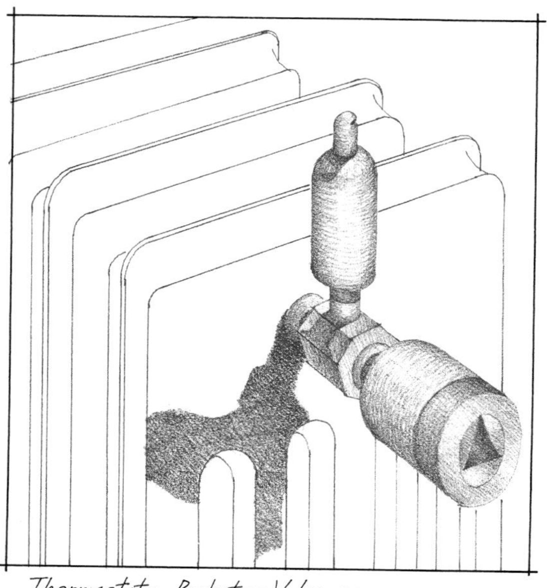

Thermostatic Radiator Valve on Steam Radiator

Thermostatic Radiator Valve on Hot Water Radiator

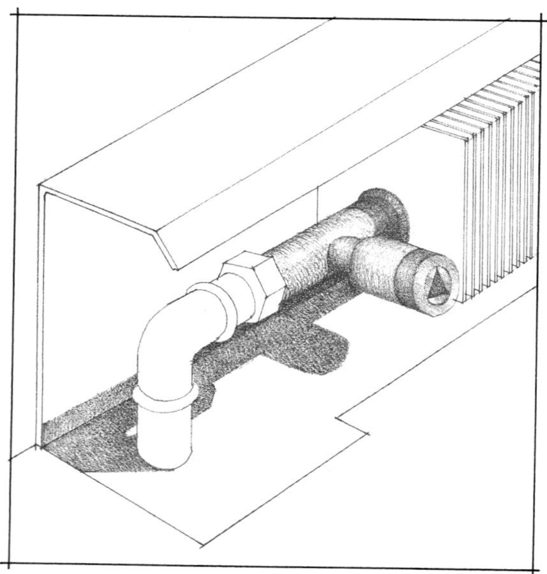

Thermostatic Radiator Valve on Baseboard Radiator

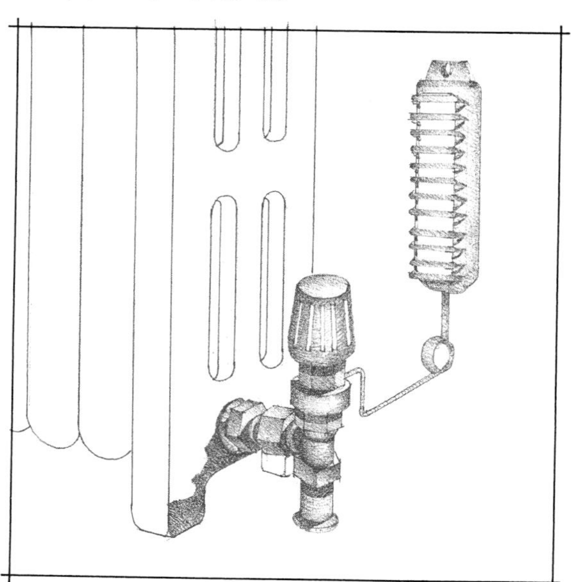

Freestanding Radiator

steam system for the cost of the valves alone ($35 to $75 each). Professional installation is recommended for other types of hydronic systems, at a total installed cost of about $200 per radiator. Some trustworthy brands are Danfoss, Braukman, Taco, Ista, and Oventrop.

Another frequent problem with hot water systems is that they're slow to respond to changing demands. When winter temperatures plunge and the house starts to lose more heat through its walls and windows, the boiler has no way of knowing that it's going to have to supply more heat — until hours later, when Aunt Ida gets

Trethewey's Tips: Insurance for Your Heating and Cooling System?

The manufacturer's warranty is your first line of protection if your new furnace, boiler, heat pump, or air conditioner gets sick on the job. Unfortunately, a lot of warranties are long on marketing hype and short on follow-through. Some are chock full of "weasel clauses" that make it practically impossible to win a claim. ("Be sure to send in the warranty registration card within 30 minutes of purchase or warranty is voided. . . . We are not responsible for equipment failures due to acts of God or ordinary wear. . . . All claims must be accompanied by an affidavit and complete set of dental records. . . .")

It's ironic that manufacturers sometimes prop up a poorly engineered system with what appears to be the longest and strongest warranty on earth. A good warranty is still worth shopping for, of course, but it ought to be just one of many criteria. And remember: The warranty is only as good as the reputation, financial strength, and good faith of the company that stands behind it. Some quality companies are so concerned about their reputation that they've been known to honor a warranty *after* its expiration date had passed.

Your second line of defense if something goes wrong is a service or maintenance contract. Like all forms of insurance, this is something of a betting game. You pay your premium to the company — usually the dealer that sold you the system — and hope that the equipment gets sick enough at some point or other to have justified your outlay. The company, on the other hand, is betting that nothing serious goes wrong with the equipment, so that it can pocket the profits.

Some contracts are written to include an annual service call, which reduces the likelihood of a major breakdown and keeps your heating system in tiptop shape. Other contracts don't include the regular service, but provide free parts and labor if there's a problem.

Whether or not it's in your interest to purchase one of these plans depends a great deal on the type and quality of the system you've bought, the reputation of the dealer who offers the plan, and the plan's specific cost and benefits.

The decision also hinges on what type of person you are. If you're going to be conscientious about maintaining the system — perhaps doing much of the maintenance work yourself — I'd suggest you put the money you would have spent on the service contract into an interest-bearing account and call it your "HVAC Rainy Day Fund." In other words, you go ahead and take the gamble, but hedge your bet by treating the equipment like royalty.

If, on the other hand, you're the sort of person who's lazy about changing air filters and other types of routine maintenance — and aren't willing or able to do any of it yourself — I'd go ahead and pay for a good maintenance contract that includes the annual service call.

the shivers and turns up the thermostat. Only then does the boiler begin to respond. Conversely, when there's a warm spell, the boiler keeps right on firing until Uncle Harry breaks out in a sweat and turns the thermostat down.

The solution is to equip hot water heating systems with a weather-responsive control. An outdoor sensor, working in tandem with the aquastat on the boiler, enables the system to anticipate rising or falling demand for heat.

The grand old Victorian we did in Jamaica Plain back in 1985 was one of several "This Old House" projects that have been retrofitted with weather-responsive controls. We located the outdoor-temperature sensor on the north side of the house, above the snow line and out of the sun. This was connected to the aquastat on the boiler, so that the temperature of the boiler water rose or fell inversely to the temperature outdoors. At 0°F, for example, the boiler works hard, sending 180°F water to the radiators. But if it warms up outside to 30°F, the boiler slacks off, circulating water at around 135°F. It's almost magic when you think about it: faster response, more comfort, less fuel. Kind of like the cruise control on a car.

Weather-responsive controls (also known as "modulating aquastats" or "reset controls") cost $250 to $500 to install. In addition to making you more comfortable, a weather-responsive control will save you up to 25 percent on your fuel bill — even more if you use it with a setback thermostat and thermostatic radiator valves.

Over the years I've used several different types of weather-responsive controls, including Centra, Tekmar, Danfoss, Honeywell, and Ista — all with good results. Your key problem isn't going to be so much in finding good equipment as it will be in locating an HVAC contractor who really understands electronic controls.

As you review the list of options that I've put together on page 133, it's important to remember that the energy savings are *not* cumulative. Since each house and heating system is unique, I can't tell you from here which modifications, or combination of modifications, would suit you best. Here's where you need to count on the experience and integrity of a good, local contractor.

Of course, it may be that your heating system is beyond the point of patchwork. If your boiler is more than 20 years old, I'd be slow to invest much money in it. You may decide that it makes much more sense to buy a new one.

Going Shopping

As with furnaces and heat pumps, boilers are tested and rated by the federal government. Each model is assigned an Annual Fuel Utilization Efficiency rating (AFUE) that estimates the boiler's annual performance. New boilers have AFUEs, that is, efficiencies, running from 80 on the low end to about 90 percent on the high.

I generally recommend oil-fired boilers with AFUEs in the 85–87 range and gas boilers rated between 80 and 85.

However, if you have a fairly large heating and hot water load, it might be worthwhile for you to pay a little extra for a higher efficiency *condensing* model. These are designed to extract more heat from the fuel, cooling the combustion gases to the point where water condenses in the heat exchanger and flue. Such systems are equipped with a drain to handle the slightly acidic condensate.

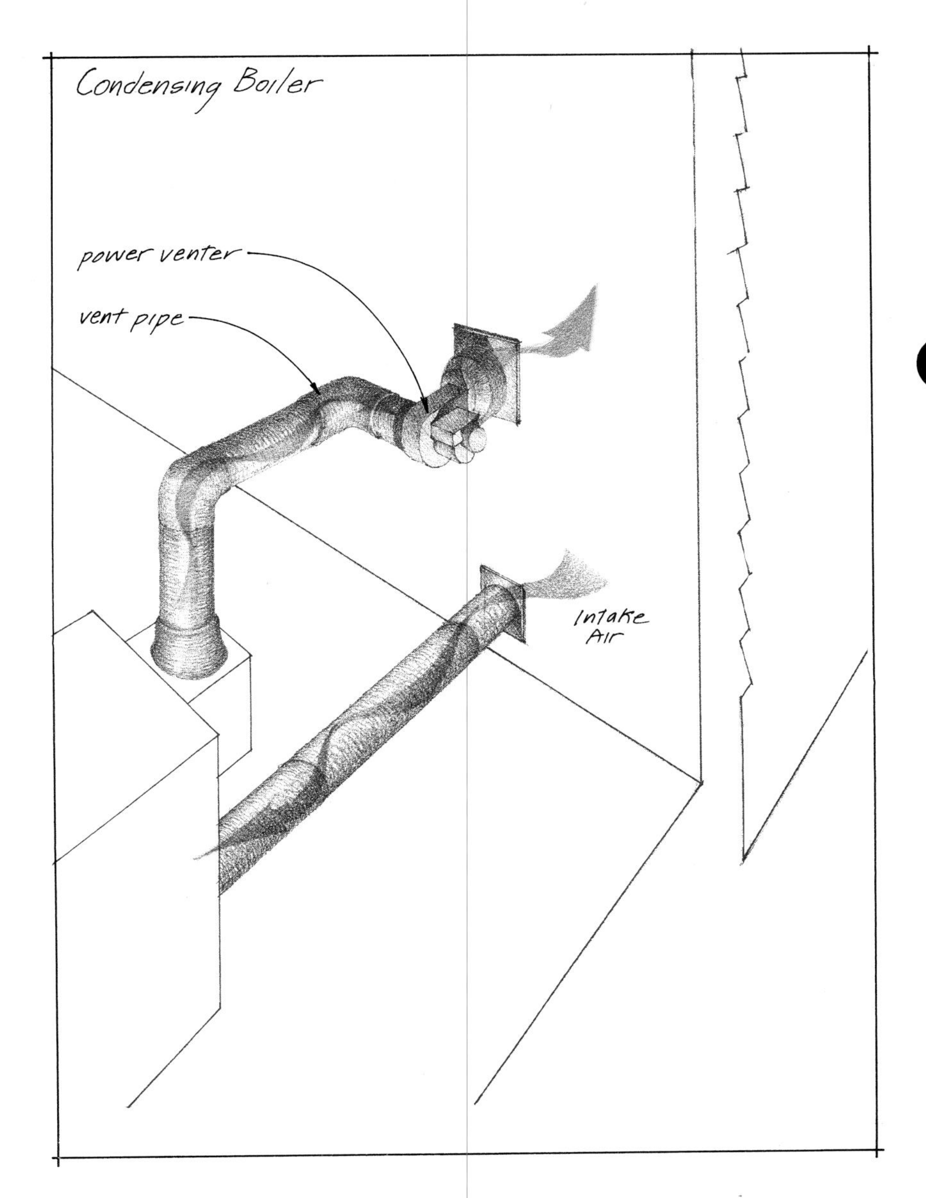

Condensing boilers have a pretty good track record for dependability — better than for condensing furnaces. As I explained in chapter 6, one of my primary concerns is with the flue. Whenever you're changing an old boiler for a new one — particularly a high-efficiency model — make sure that the existing chimney or flue is adequate. Condensate can chew up the mortar in a chimney, weakening its structure and creating a fire hazard.

Depending on the temperature of the flue gases coming out of the boiler, it may be possible to use plastic flue pipe, such as Plexvent or Ultravent, which is piped out through a side wall, up through the roof, or up the middle of the existing chimney. Stainless steel flue pipe, though expensive, is another option.

Apart from making sure that the boiler and flue are safe to operate, nothing is as important as proper sizing. As with furnaces and heat pumps, the HVAC industry has a tradition of grossly oversizing boilers. The results: high cost, discomfort, and needless wear and tear on the equipment.

Make sure that your contractor does a proper heat-loss calculation on your house and that the boiler you select is sized to meet the load — not exceed it. A properly sized boiler should run almost continuously in very cold weather.

Remember that in the process of installing the new boiler you may be effecting changes in your house that you didn't expect.

For example: Replacing an old aspirating boiler with a closed-combustion model will probably reduce the amount of natural ventilation your house gets. (By drawing its combustion air from inside the house, the old boiler depressurized the house and coincidentally increased the infiltration of fresh air.) That's why I always recommend that the home's ventilation needs be reexamined whenever new heating equipment is installed.

A new boiler may also pose an indirect threat to water pipes and other equipment. I've lost count of the number of times I've been called in to fix burst water pipes that had frozen in the area around a new boiler. As long as the inefficient old boiler was sitting there, losing heat through its jacket to the surrounding space, there was enough incidental heat to keep the pipes from freezing. Then along came the new, superinsulated, high-efficiency boiler, and suddenly there wasn't enough heat leaking out to protect the pipes. If you're changing an old boiler that's located in an unheated basement, surrounded by pipes and other things that might be damaged by a freeze, take heed.

For reasons that I mentioned before, steam boilers aren't recommended for new installations. But the question remains: What do you do if you own a steam system and the old boiler dies?

The "This Old House" crew has tackled aged steam systems four times since the show began, including our very recent work on Chris and Joan Hagger's house in Wayland, Massachusetts.

As you may recall from those shows, we junked the Haggers' old oil-fired steam boiler (aided by some timely grunt-and-tote work from Steve Thomas) so that we could put in a new gas-fired hot water boiler with a circulator. All of the radiators, which were originally one-pipe steam, had to be fitted with return lines to get the circulating water back to the boiler. The best solution was to snake PEX plastic tubing down through the walls — the way an electrician snakes wire — so that we could connect each radiator to the return side of the boiler. Because we didn't want to take any chances with leaks in the walls, we used the same superplastic tubing that I recommend for in-floor heating: cross-linked polyethylene.

The Relative Cost of Energy

Energy Type	Cost per Unit of Fuel	Price per Million BTU
Electricity	8.0 cents per kilowatt hour	$23.45
Propane	66.2 cents per gallon	$ 7.25
No. 2 heating oil	93.4 cents per gallon	$ 6.73
Kerosene	78.6 cents per gallon	$ 5.82
Natural gas	$5.85 per thousand cubic feet	$ 5.67
Dry hardwood	$80.00 a cord	$ 3.33

Note: Prices will vary region to region. Other factors, such as the fuel's availability and the efficiency of heating equipment, should be taken into account when selecting a heating fuel.

SOURCE: Department of Energy. All prices are 1992 national averages, except electricity, which is a 1991 average, and dry hardwood, which is a 1992 spot price in New Hampshire, cut, split, and delivered.

How We Heat Our Homes

Fuel	Millions of Single-Family Homes	Percentage of Total
Natural gas	36.7	57%
Electricity	12.2	19%
Fuel Oil	7.4	12%
Wood	3.6	6%
LPG	3.2	5%
Kerosene	.5	1%
Other	.3	1%
None	.5	1%

SOURCE: U.S. Department of Energy, Energy Information Administration. From "Housing Characteristics 1990." Data include a total of 64.4 million single-family residences. Total percentage exceeds 100 due to rounding.

In addition to the new boiler and the superplastic return lines we installed, each radiator was equipped with a new thermostatic valve. *Voilà!* The Haggers' old house was zoned!

If you're considering a steam-to-hot-water conversion, be forewarned that the change will drop the heat output of your radiators by as much as 30 percent. We were able to make the change on the Wayland house because we tightened and reinsulated the house first, which cut the heating load.

The Wayland house was only one of many projects in which we changed oil-fired equipment for natural gas. But that doesn't mean that I'm unequivocally recommending natural gas for all new equipment. My instincts tell me that it's usually best for homeowners to stick with the heating fuel they're already using. The cost of running in a new gas line or installing fuel oil or propane tanks puts you in a hole right from the start. If you're lucky enough to be able to choose between several fuels, the choice gets tougher. To avoid getting hate mail from fuel oil dealers, the American Gas Association, or the Edison Electric Institute, I've tried to be objective in presenting the pros and cons of different heating fuels on pages 140–141.

Which Fuel for the Future?

People are always asking me, "Richard, what's the best heating fuel, oil or natural gas?"

Being an artful dodger, I always smile politely and answer, "Yes."

The projects we've done on "This Old House" have tended to favor natural gas over fuel oil for a couple of different reasons. First, natural gas furnaces and boilers generally burn cleaner, both in the amount of soot they deposit in the combustion chamber and in the amount and type of pollutants they send up the flue.

Having said that, I must point out that it's not as true as it used to be. Some modern oil equipment, like the Viessmann boiler I demonstrated on the show, burns so hot that it works like a self-cleaning oven. Unlike conventional oil boilers and furnaces, which can soot up over the course of a winter and gradually lose their efficiency, some of the state-of-the-art oil-fired equipment is every bit as clean-burning as natural gas.

A small, additional advantage to using natural gas in your home is that you don't have to install a storage tank on the property. At Chris and Joan Hagger's house, in Wayland, Massachusetts, we won quite a bit of extra space in the basement by removing the old fuel oil tanks and using gas instead.

Although I can't predict the long-range price trends of different fuels, I do know that natural gas reserves in the United States are much more plentiful than oil, and would speculate that the price of natural gas is going to reflect that relative abundance over time. The closer you get to the gas fields in Louisiana, Texas, Oklahoma, Kansas, Colorado, and Wyoming, the more reliable and less expensive the gas supply becomes.

While it's true that natural gas supplies here in New England have occasionally fallen short, and the price advantage over oil hasn't always been apparent, the completion of the Iroquois, Trans-Canadian, and other new pipelines — linking the Northeast to the rich gas fields of western Canada — should make plenty of natural gas available at competitive prices.

As for overall safety, I think oil has a slight advantage. While the risks of explosion or asphyxiation are quite small with central gas heating systems, they can and do occasionally go boom or cause a death from carbon monoxide poisoning. I'll discuss the issue of safety in chapter 8.

Compared to the price of fuel oil or natural gas, electricity is the Cadillac of fuels. At 8 cents a kilowatt hour it's equivalent in heat value to oil priced at $140 a barrel! I wouldn't generally recommend electricity for heating, with the following exceptions:

- For very tightly built and well-insulated homes where the heating load is minuscule.
- For homes in the deep South or summer homes that have only a few heating days a year.
- For people who want to consolidate their heating and cooling in one system — that is, an electric heat pump. (While air-to-air heat pumps can function anywhere in the United States, ground-coupled heat pumps generally offer better performance, especially in cold-weather climates.)

Now, here are five additional tips on buying a new boiler:

- Beware of boilers that use a very small volume of water. These designs were developed back in the 1970s as a response to the energy crisis, but experience has shown that they cycle on and off so often — up to 40,000 times a heating season! — that the equipment fatigues. A desirable cycling time for a hot water boiler might be five to six minutes (never two or three).
- Consider models with sealed combustion. For the various reasons I spelled out in chapter 6, the trend is to isolate the boiler from the house. This is accomplished by providing the combustion chamber with its own fresh-air vent and using fans to control the intake and exhaust.
- Design the installation so that the boiler also heats your

- In those few remaining utility districts where electricity is still selling at a nickel per kilowatt hour. (Warning: it may not stay that cheap forever!)

There are other fuel sources, of course. In northern New England and some other parts of the U.S., wood is the cheapest fuel around. We'll talk about wood stoves and fireplaces in chapter 8.

Propane, which is liquefied natural gas, is becoming increasing popular in some (especially rural) parts of the country. Though propane is generally more expensive than natural gas and requires an on-site storage tank, it's usually less expensive — both for space heating and hot water — than electricity.

As far as the environmental impact of different fuels is concerned, there are no easy answers. Oil, gas, propane, and wood all release environmentally damaging combustion by-products, including nitrogen oxide, sulfur dioxide (oil), carbon monoxide, and carbon dioxide.

Electricity isn't any better. The people who promote electric heating love to talk about how clean and efficient it is for use at home, while a few miles away, down at the generating plant, they're burning tons and tons of coal or fuel oil or garbage, or reacting uranium into too-hot-to-handle radioactive waste.

Hopefully, the various solar and geothermal technologies being developed will provide some cleaner alternatives — but of course, they aren't perfectly perfect either.

I'll leave you here with some additional tips on choosing fuels.

- Don't change heating equipment that's still usable simply because you want to change fuels. Wait till it's time to replace the system, then consider switching.
- Don't be overly influenced by day-to-day pricing or political events. Just before the war with Iraq broke out, the price of fuel oil shot up to $1.29 a gallon (versus a natural gas equivalent of 87 cents). But within weeks the price of fuel oil was headed south again. Today I see it quoted in the *Boston Globe* at 85 cents a gallon. Remember: Your new heating system is going to last you well beyond the turn of the century, so you need to be thinking *long term* when it comes to fuel.
- As a rule of thumb, you'll save money sticking with the fuel you've already got rather than putting in a new gas line, fuel oil tank, or propane tank. The exception is electric resistance heating, which can almost always be replaced with another fuel at a good savings.
- If you have the luxury of choosing between several available fuels, talk it over with your contractor. The key factors are: long-term availability of the fuel; local prices (history and projected); short- and long-term incentives from the supplier; service from the supplier; and the type and efficiency of the heating equipment you're considering.

domestic water. This is simply done by running a hot water pipe out of the boiler and through a heat exchanger inside a superinsulated water storage tank. (The old hot water tank is mothballed.) By getting all your space heating and hot water from one efficient boiler, with only one fuel line, one burner, and one flue, you can chalk up big savings. (More on this in chapter 9.)

- Opt for the best warranty you can find. Two years on parts and 20 on the heat exchanger is the industry standard.
- Stick with reputable manufacturers who have a track record. Some trustworthy names in the industry include Weil-McLain, Burnham, Hydrotherm, Slant Fin, and Viessmann.

●

Direct Heating Systems

A surprising number of people in the United States depend on direct heating for their wintertime comfort. By that I mean, their heat source has no ductwork or pipes to distribute heat around the house.

Included in this category are such diverse heating types as baseboard electric resistance heating, radiant electric, wood stoves, wood and gas fireplaces, pipeless wall and floor furnaces, and space heaters.

After crunching some numbers that I borrowed from the U.S. Department of Energy, I discovered that there are more than 14 million single-family homes — that's more than one out of every five — that rely on some type of direct heating.

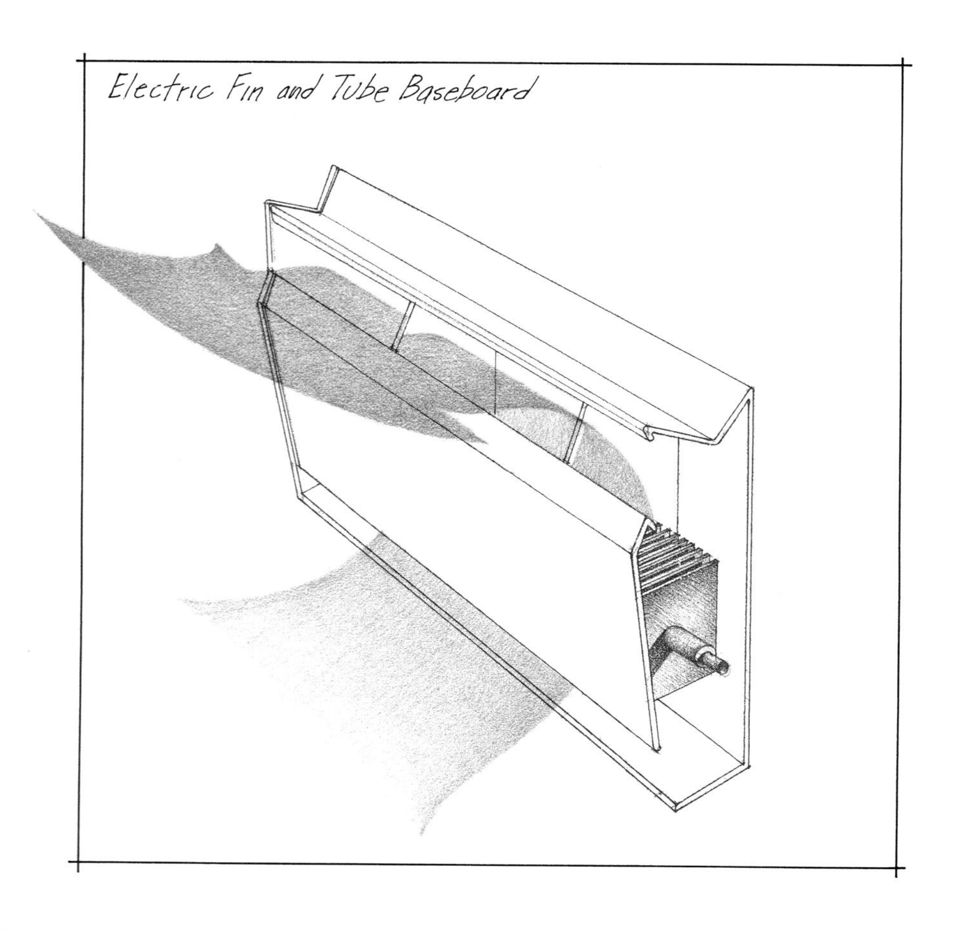

Baseboard Electric

Baseboard electric radiators are the essense of simplicity. Pop the cover off the radiator and you'll see the heating element — wrapped in a tubular casing — with metal fins positioned along its length. When electricity flows into the element, it gradually heats up — the same way the elements in a toaster or electric oven produce heat.

Electric baseboard radiators come in one- to twelve-foot lengths, with ratings of 100 to 400 watts per foot, so they're very flexible in meeting different space and load requirements. Like their hydronic cousins, electric baseboards are usually placed around the outside walls of the room and positioned under windows to counteract the infiltration, conductive heat losses, and drafts associated with the window glass and frame. As with copper-finned hydronic radiators, the fins on electric models serve to increase the heated surface area and promote the movement of air up through the radiator.

No other heating system I can think of offers as many *apparent* advantages as baseboard electric resistance heating.

To begin with, it costs less to install than any of its rivals, with the possible exception of a wood stove. That's why builders trying to drive down the first cost of their homes order these systems by the mile.

Second, electric baseboard radiators are easy to zone. Some models have a thermostatic control built right into the radiator; others are designed to operate with wall-mounted thermostats. If you want to heat your bedroom and let the living room cool down — or vice versa — electric baseboards with room-by-room thermostats make it easy.

Third, you avoid all of the costs and possible problems associated with having a combustion heater in your house: With baseboard electric there's no need for a chimney or flue, no worries about backdrafting combustion gases or depressurization, and no need to run in a natural gas line or install a fuel oil tank. All of these costs and concerns are transferred to a distant site, where the utility runs its electric generating plant.

Electric radiators are fairly quiet too, though the element and metal housing tend to crank and rattle a little as the unit heats up and cools down.

The big bugaboo with baseboard electric, of course, is its operating cost, which is so onerous in some utility districts that it outweighs all of the advantages — and then some.

Had nuclear power delivered on its promise to produce electricity "too cheap to meter," I might be able to recommend

A Tale of Twenty States
(Average Cost of Residential Electricity in Cents per Kilowatt Hour)

The Budget Busters . . .		. . . and Bargains	
1. New York	12.0	1. Washington	4.4
2. Rhode Island	11.0	2. Oregon	4.8
3. California	10.8	3. Idaho	4.9
4. New Jersey	10.8	4. Tennessee	5.6
5. Alaska	10.7	5. Kentucky	5.7
6. Connecticut	10.5	6. Montana	5.8
7. Hawaii	10.5	7. Nevada	5.9
8. Maine	10.5	8. West Virginia	5.9
9. Massachusetts	10.4	9. Wyoming	6.0
10. New Hampshire	10.4	10. Nebraska	6.1

Average for 50 States and Washington, DC: 8.0
SOURCE: U.S. Department of Energy, Energy Information Administration. "Electric Power Annual, 1991."

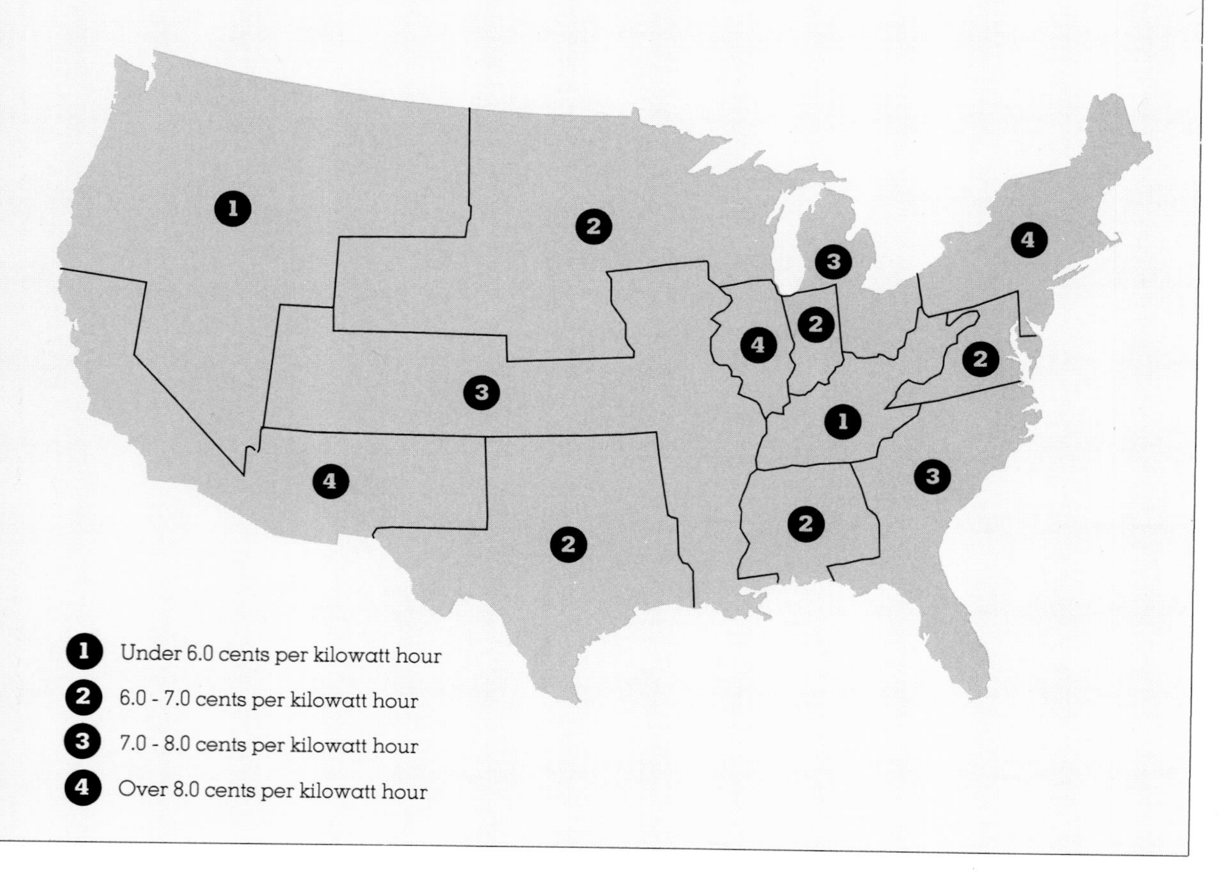

electric resistance heating. As it is, there are a lot of homeowners in New England and elsewhere in the United States who bought or built a home with baseboard electric, only to see their wintertime electric bills soar through the stratosphere.

And while it's true that all of the fuel loading, combustion, and pollution associated with electric heating take place at a distant (or not-so-distant) generating plant, the economic and environmental costs still have to be paid. These costs are partly reflected in everyone's electric bill and partly absorbed as "societal costs," that is, through the taxes we pay to subsidize and regulate utilities, through rising health insurance premiums, through the money society spends to handle radioactive wastes and combat air, ground, and water pollution, and through other "invisible" costs.

Unless you live in a climate where the winter is mild, or have cheap electric rates, or live in a superinsulated house with a peanut-size heating load, you should look for something other than baseboard electric as your primary heating system.

Radiant Electric

In the event you do decide to use electric resistance heating, you may want to consider a radiant system. Instead of using baseboard radiators to convectively heat the air inside the room, radiant electric systems are designed to heat objects only.

One way to deliver radiant electric heat to a room is to install flexible element heaters above the ceiling or underneath the floor. These prefabricated sheets have a metallic or carbon-graphite conductor that's sandwiched between two layers of insulating plastic or, in some cases, printed directly on the plastic. As the element heats up, warmth spreads into the sheetrock or ceiling tile (in overhead installations) or into the plywood subfloor and flooring (in underfloor installations). Once the finish is warmed, it radiates heat to various objects inside the room, the most important of which, of course, is *you*.

One of the first steps in a proper installation is to insulate the cavity between the joists so that as much heat as possible will be directed into the room. Next, the flexible element heater is stapled into place and wired into the home's electrical system. Last, the ceiling or subfloor goes in, sealing the cavity with an intervening air space.

Some brands of flexible element heaters are pre-cut at the factory while others are delivered in rolls so that they can be cut to fit at the job site. Widths range up to 48 inches. Three of the leading brands are Flex-Watt, Aztec Flexel, and Flex-Heat.

Another approach to radiant electric heating — for ceilings only — is to use a special type of gypsum board that has nichrome heater wires embedded in its core. Known under the brand names Panelectric and Suncomfort, gypsum board heaters are a lot heavier and less flexible to install than plastic sheets. In fact, the ceiling has to be carefully planned in advance so that correctly sized panels can be ordered. On the plus side, gypsum board heaters are relatively inexpensive to buy and save material and labor by doing double duty as the finished ceiling.

A third approach to radiant heating — though not a common one — is to install a grid of specially sheathed electrical cable in the floor. The cables can be embedded directly in the concrete or masonry when the floor is poured or laid, much the same way that radiant hydronic systems are installed. Raychem Corporation, which makes RaySol in-floor heating systems, uses a special polymer in its electric cables that enables the system to regulate itself.

All of the radiant electric systems I've described above are most cost-effectively installed during new construction work or as part of a major rehab — that's when the ceiling and floor cavities of the house are open and easily accessible to an electrician.

But radiant electric heat can also be installed using modular panels, which can be mounted flush against an existing wall or ceiling. The panels come in various sizes, wattages, and colors (some can be painted) so they can blend in with the decor. Though modular panels tend to be pricey compared to other radiant electric systems, they can be retrofitted without tearing into the room's finish. And since modular panels don't have to heat up the sheetrock or ceiling tile before they begin to deliver heat, they respond a lot faster to a call for heat. Enerjoy and Aztec are a couple of the better-known brands.

In my view, radiant electric heat has three big advantages: it's easy to zone, it doesn't take up any floor space for radiators, and there's no maintenance.

In certain types of applications, such as in rooms that are only intermittently heated, rooms with very high heat loss, and for heating a small part of a large room, radiant electric systems can cost less to operate than electric baseboard radiators and other warm-air systems.

In more typical situations, though, the savings are relatively small or nonexistent. So beware of wildly exaggerated claims that radiant electric will save you 30, 40, or 50 percent compared to other systems. (Any day now I'm expecting one of these outfits to claim 100 percent energy savings. Maybe even 110!) In any case, ask your contractor to give you a clear estimate of what the operating costs are going to be *before* you go ahead. If savings are promised, be sure to get them in writing. Other key points to consider when shopping for radiant electric are the total installed cost of the system and its response time.

When Electric Makes Sense

One place where electric baseboard radiators or radiant electric heat can make good sense is in a new room addition where the cost of expanding the home's existing forced air or hydronic heating sys-

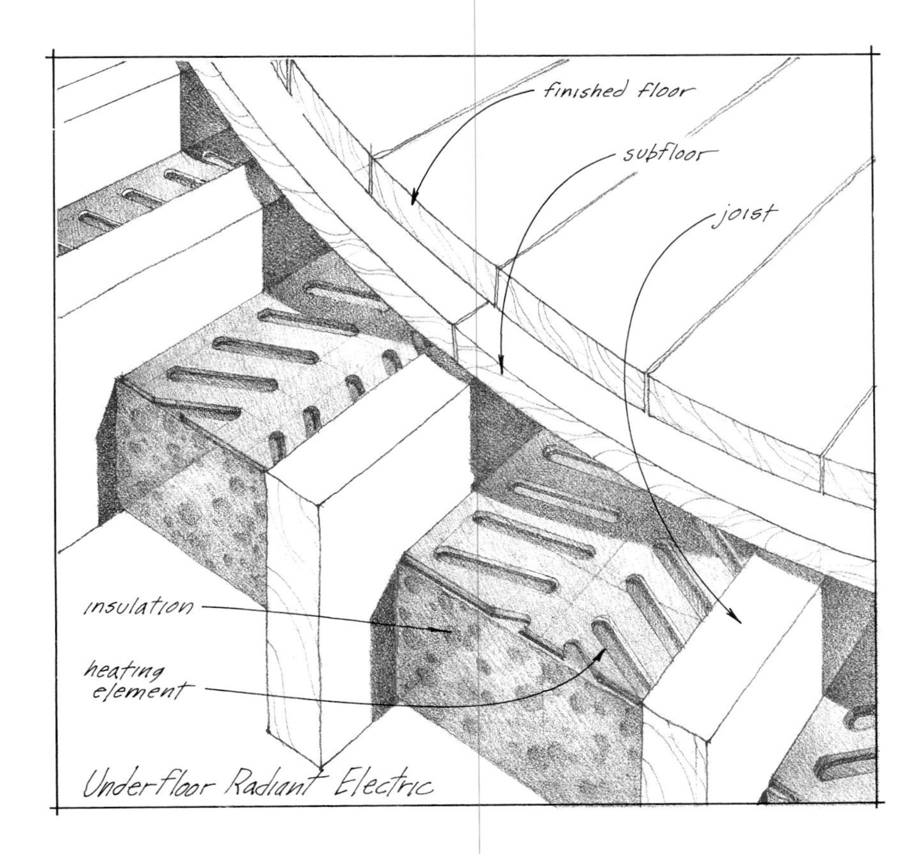

tem is so high that it makes the electric — operating costs included — look cheap.

Another cost-effective use for electric resistance heating might be as a back-up system for people who use wood or coal stoves as their primary heating source. When no one's at home to stoke the fire, the electric automatically kicks in, keeping the house just warm enough to protect the plants and pipes. In similar fashion, electric heat could play a supporting role in passive solar homes.

An electric baseboard radiator or radiant panel might also be a good way to warm up a cold spot in your home, as opposed to using some other type of gas or electric space heater.

Maintenance duties are few and far between with electric resistance heating, which is probably its strongest advantage. Baseboard radiators should be cleaned occasionally with a brush or vacuum. Drapes, rugs, furniture, and electrical cords need to have a few inches of clearance, both as a safety precaution and to let room air flow freely up through the radiator. As for radiant electric systems, there's no maintenance at all.

In every case I would recommend that you get a licensed electrician to handle the installation. Your home's existing circuits may not be able to handle the increased amperage required by electric resistance heating.

As a safety precaution, the system should have a limit switch that automatically breaks the circuit if the unit starts to overheat. It goes without saying that the equipment you use should be approved by Underwriters Labs, ETL Testing Laboratories, or other accredited test lab.

One final word about safety: Both baseboard electric radiators and radiant electric systems generate electromagnetic fields — or EMFs — inside the home. Though the science is very immature and the evidence sketchy at this point, researchers believe there is a link between EMFs and certain types of cancer. In the home, they cite TV and computer screens, clock radios positioned near the bed, hair dryers, malfunctioning microwaves, and electric blankets (particularly for pregnant women and children) as sources of concern.

Until the health effects of EMFs are better understood, some researchers and health officials are recommending a policy of "prudent avoidance." I have to wonder exactly what that means, since all of us are in effect swimming through a sea of electromagnetic fields every day of our lives, including the planet-size EMF that enfolds the earth itself.

Does "prudent avoidance" mean, for example, that we should abandon electric resistance heating altogether?

Ed Mantiply, a scientist with the U.S. Environmental Protection Agency, says there's no way at this time to assess precisely the risks associated with individual appliances or heating systems, because the mechanisms by which EMFs cause biological changes in humans are not well understood. Early research suggests that the strength and proximity of the magnetic field are key variables, which is why electric blankets have come under close scrutiny and are in fact being redesigned by some manufacturers to reduce any possible hazard.

The potential risks associated with baseboard electric or radiant electric heating systems would depend on the amount of current (amps) running through the system, how close you are to the field and how often you're exposed to it, and the geometry of the wiring (systems with big open loops set up a stronger magnetic field than do systems that route current back along the same line). Other variables, such as how often the system cycles on and off, may also turn out to be important.

If you're thinking about installing an electric resistance heating system and are concerned about the possible health effects on you or your family, I suggest you talk the matter over with your contractor and give a ring to your state health office. If you're willing to spring for a stamp or long distance call, the U.S. Environmental Protection Agency is probably your best source for up-to-date information. You can call their Public Information Center in Washington, DC, at (202) 260–2080 or write: U.S. Environmental Protection Agency, Office of Radiation Programs (ANR-461), Washington, DC 20460.

Another good source for information is Carnegie Mellon University, which offers a booklet titled *Electric and Magnetic Fields from 60 Hertz Power: What do we know about possible health risks?* (Send $3.50 to the Department of Engineering and Public Policy, Carnegie Mellon University, Pittsburgh, PA 15213.)

Wood Stoves

A good friend of mine who lives up in New Hampshire, and for years has heated his home with wood, insists that there's no better fuel in the world.

"Wood's the only fuel around that can heat you *five* times," he points out. "It heats you once when you have to cut it. Then again when you split it. A third time when you go to stack it. A fourth when you tote it into the house. And a fifth and final time when you finally get the fire going!"

Actually, I think my friend is forgetting a *sixth* possible heating — the one he gets when he has to clean out the ashes.

Without belaboring the point, I think it's fair to say that wood heat is the most labor-intensive way you can choose to heat your home. Even if you buy cord wood that's cut, split, and delivered, you'll still find yourself doing a lot of stacking, lugging, and stove-tending.

Having said that, I want to add quickly that wood heat has some powerful advantages, which is why more than four and a half million homeowners use it as their primary heating source. (Millions more use wood stoves and fireplaces as room heaters or to complement an oil, gas, or electric system.)

In many parts of the United States, including northern New England, the Pacific Northwest, and parts of the South, wood costs less per million BTUs than any other fuel around. (A BTU, short for "British thermal unit," is the amount of heat required to raise the temperature of a pound of water by one degree Fahrenheit. To put it another way, a BTU is roughly the amount of heat released in burning a wooden match.) As shown in the chart on page 139, other fuels can cost up to eight times as much as wood does.

Wood has the further advantage of being renewable — that is, it's naturally replaced year in and year out — while the oil and natural gas we use comes from finite and declining reserves. And wood is a home-grown source of energy, abundant and secure. No need to rely on tankers from Saudi Arabia or pipelines winding their

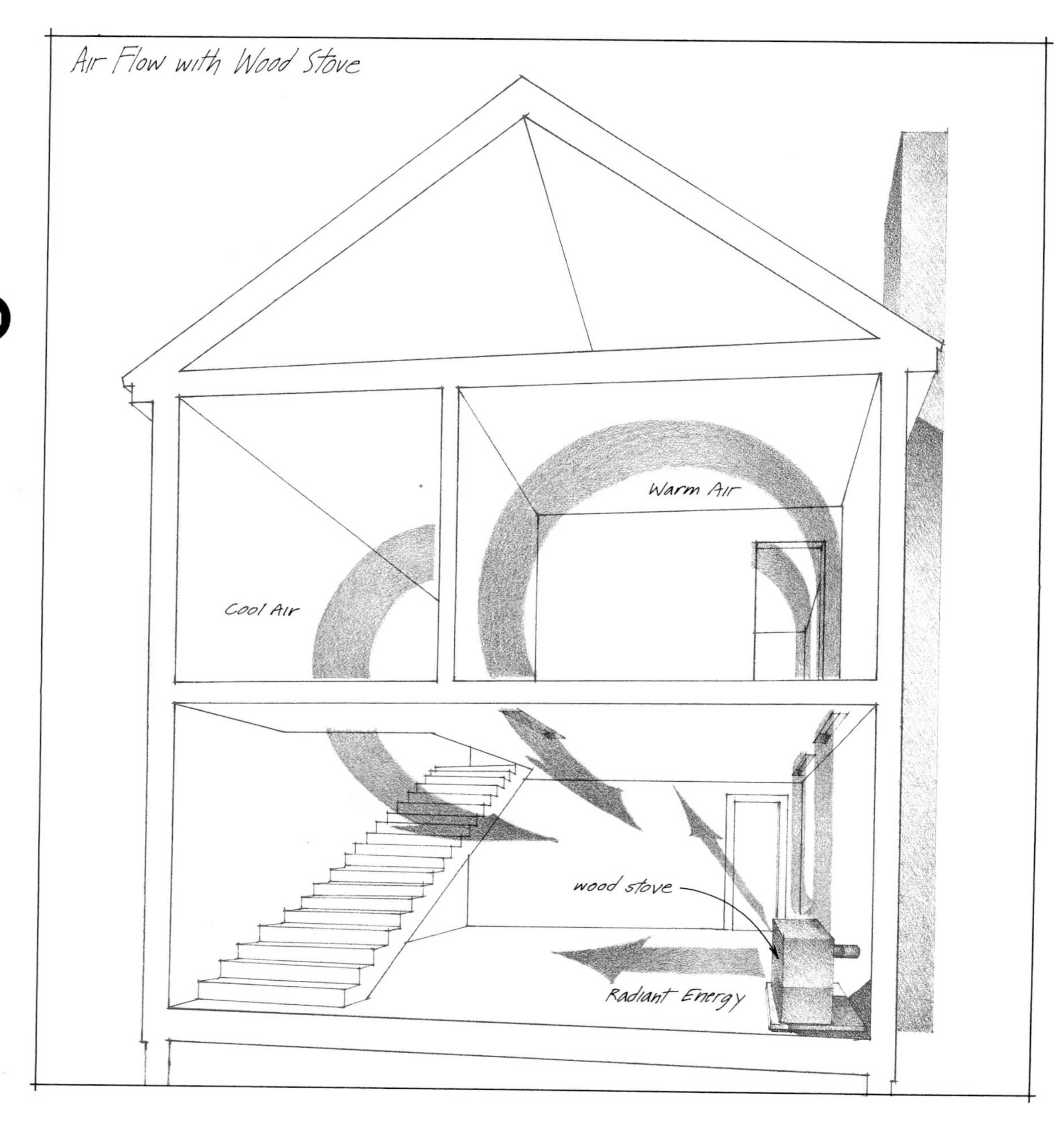

way down from Canada. And no damage done to our nation's balance of trade either.

Last but not least, there's something fundamentally appealing about a wood fire, something that dates back to our earliest ancestors, something buried in our psyches that loves the radiant warmth and fiery blaze.

As far as the heating efficiency of wood stoves is concerned, I've got some bad news and some good.

The bad news is that old-style wood stoves — including most models built before 1985 — are only 20 to 60 percent efficient, depending on the type of stove and how it's operated. In other words, 40 to 80 percent of the energy available in the wood is going right up the chimney as wasted heat and pollution.

It comes as a rude surprise to a lot of homeowners, particularly those who bought their stoves thinking they were making a positive contribution to the environment, that wood stoves can be a serious source of air pollution. Yet the evidence is clear: Studies conducted by Monsanto Corporation and other researchers have found that wood smoke contains large amounts of polycyclic organic matter — a known carcinogen — as well as particulate matter, formaldehyde, carbon monoxide, and a host of other unsavory substances. So much for the "invigorating" smell of a wood fire!

As more and more people turned to wood heat in the late 1970s and early '80s, health officials began to worry about the effects of declining air quality, particularly on the aged and the very young. These problems have been especially serious in valley communities where a large percentage of homeowners heat with wood.

For example, wood smoke pollution got so bad in the tiny resort community of Telluride, Colorado, the town fathers (and mothers too, presumably) had to budget $144,000 in rebates to encourage citizens to buy new, clean-burning stoves or switch to other types of fuel. Other towns and cities that have been plagued with wood smoke problems include Missoula, Montana; Mammoth Lakes, California; Vail, Colorado; and Juneau, Alaska. Each of these communities has had to use a combination of penalties and incentives to address the problem.

Which is where the good news comes in.

A new environmental law, hammered out between wood stove manufacturers and the U.S. Environmental Protection Agency, places strict emission limits on all new wood stoves sold in the U.S. Since the law took effect in 1988, a new generation of cleaner, more efficient stoves has entered the market, and the nation's wood smoke problems are starting to subside.

Compared to their smoke-belching predecessors, state-of-the-art wood stoves achieve a 70 percent or better reduction in particulate emissions and are up to 80 percent efficient, rivaling the efficiency of many furnaces and boilers.

Some of these new EPA-certified stoves are equipped with catalytic combustors, which are similar to the catalytic converter in your car. When the stove reaches 500°F, special catalytic metals (platinum or palladium) are activated inside the combustor, enabling the stove to burn wood gases and smoke that would otherwise go up the flue. The results: More heat for your home and less pollution in the atmosphere.

Other types of wood stoves meet the EPA standards without

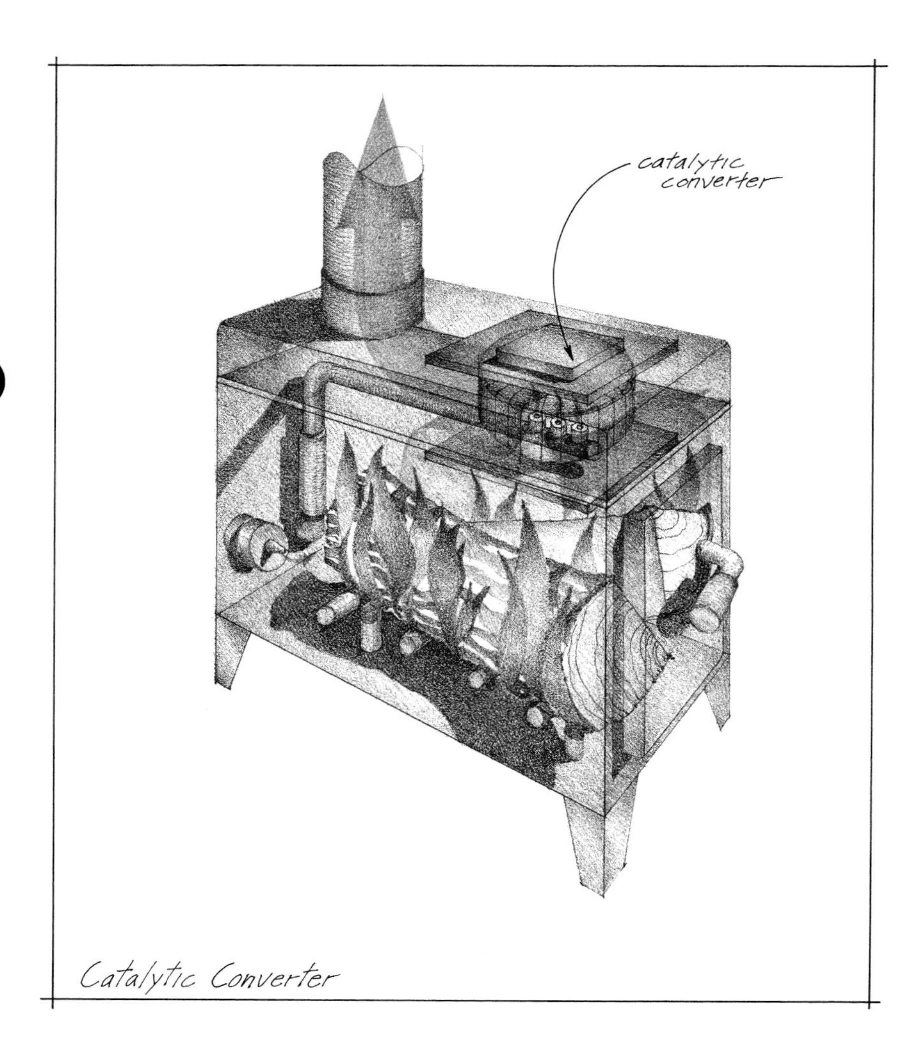

Catalytic Converter

using catalytic combustors, by routing combustion gases and smoke back through the stove to achieve secondary combustion. This type of stove boosts efficiency and reduces pollution the same way the catalytic models do — that is, by burning gases and smoke inside the stove rather than releasing them up the flue.

The argument for buying one of these new EPA-certified stoves — even though your old stove works fine — becomes pretty persuasive when you consider the fact that a new model will reduce the amount of wood you burn by about 30 percent! So in addition to saving money (if you buy your wood), you can also expect to put 30 percent less time and energy into stacking, toting, and cleaning up ashes.

Maintenance

In addition to cleaning out the wood ashes every so often and sweeping up the hearth, wood stove owners need to keep a careful eye on the chimney. One of the nastiest by-products released in wood burning is creosote, a coal-black, carcinogenic residue that

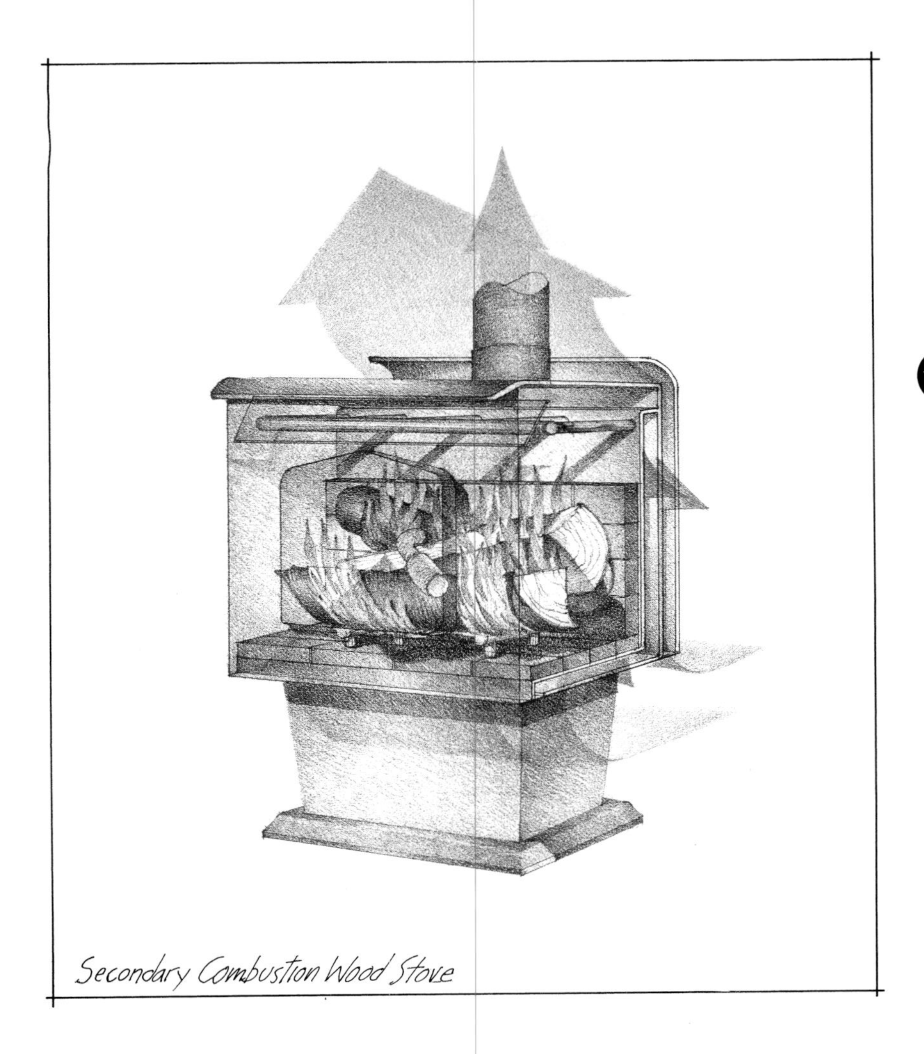

Secondary Combustion Wood Stove

builds up on the chimney liner. If the creosote isn't periodically cleaned out, it builds up to the point where it not only reduces the fire's draw, but also becomes a serious fire hazard.

The National Fire Protection Association ranks wood stoves as one of the leading causes of home fires in the United States. The main culprit, of course, is chimney fires, caused when creosote deposits ignite inside the flue.

Chimney fires can be easily avoided though, by having the chimney professionally inspected at the start of the heating season and then cleaning it as necessary to remove built-up creosote. You can buy a brush-and-rod kit and do this yourself if you like, saving the $50–$100 it would take to hire a professional sweep. (For old chimneys with questionable integrity or chimneys that have an offset flue, I'd recommend you get a pro.)

If you do decide to clean the chimney yourself, be forewarned that it's one of the worst jobs known to man. You'll need to put on protective work clothes, including a dust mask and gloves.

You may also want to "clothe" any furniture, rugs, or other valuables that are near the ash chute with plastic drop cloths or sheets so that they don't end up covered with soot.

Sweeping a chimney is a lot easier done with two people than one, so look around for a volunteer to help you. To start, one person stations himself on the roof, where he can push the wire brush down into the chimney, adding extension rods as they're needed. Before the brush goes in, through, he fastens a rope to it (threaded through an eye on its lower tip) and drops the rope down through the chimney. The second person (who lost the toss) squats at the bottom of the ash chute, receives the rope, and prepares to pull down on the brush as needed. Using this two-person push-and-pull method, you can be pretty sure that you're not going to get the brush stuck in the chimney.

Once you've finished the job, be sure to sweep up all the creosote and bag it tightly so that it can be properly disposed of. Then dump your work clothes into the wash and hit the showers pronto.

One of the hidden beauties of the new clean-burning stoves I spoke of earlier is that they reduce the formation of creosote in the chimney by as much as 90 percent. That means fewer cleanings, much less chance of a chimney fire, and eventually — as the insurance companies catch on — lower home insurance rates.

Catalytic stoves do require a little bit of extra maintenance compared to conventional and secondary combustion stoves, but not much. Here's the routine:

- The catalytic combustor should be vacuumed off or brushed clean two or three times during the heating season to remove any fly ash that's caught in the unit. It's a quick and simple job.
- A more thorough cleaning, which involves boiling the combustor in distilled water and vinegar, is recommended every one to three years depending on how frequently the stove is used. Your owner's manual will spell out how to do this. Don't try to clean the combuster with water, compressed air, or scraping tools.
- Make sure that you have a reliable temperature gauge on the stove — it's the only way you can tell if the catalytic combustor is working properly. If you aren't getting operating temperatures between 700 and 1600°F, the combustor is probably clogged or worn out. Depending on how heavily you use the stove, you can expect a catalytic combustor to last two to six years. A replacement unit, which you can install yourself, costs $100 to $200.

- Burn only natural wood. That means no painted or treated woods, artificial logs, coal, trash, lighter fluids, or chemicals. (This is good advice for *any* type of wood stove or fireplace, by the way.)

Shopping for a New Wood Stove

EPA-certified wood stoves start at about $500 and range all the way up to $2,400. With the higher-priced models you're paying for the brand name, color enameling, and ornamental details, which have nothing to do with how efficient or clean-burning the stove is.

One of your first and best clues to stove performance is the EPA certification label, which is affixed to every new wood stove on the market. As you can see in the replica, the label gives information on the stove's pollution characteristics (in grams per hour), efficiency (on a sliding scale from 50 to 100 percent), and heat output (in BTUs per hour).

Unlike boilers and furnaces, which have a single BTUs-per-hour rating, wood stoves have a high-low range (i.e., 9,000–41,300 BTU/hr.). That's because a wood stove is periodically stoked and then left to burn, so that its heat output tends to rise, hit a plateau for a while, and then fall again.

When shopping, it's important to remember that the "Maximum Heat Output" listed in the manufacturer's literature or described by the salesperson can only be maintained by more or less constant stoking. Likewise, the "Maximum Burn Time," which ranges from six to twelve hours, includes both the early-ignition stage of the fire and the dying-ember stage, neither of which throws off much heat. Long burn times are especially important to homeowners who want their stove to burn through the night and still have enough embers in the morning so that the fire won't have to be relit from scratch.

In addition to listing BTUs per hour, most wood stove manufacturers describe the capacity of their various stoves by indicating how many square feet it's capable of heating. While this is helpful in selecting the right stove for your space, bear in mind that the results are going to depend a lot on *how that square footage is configured.*

If your house is designed around an open concept, with the wood stove positioned in its center, you may not have any heat distribution problems at all. But in houses that have hallways, with distant bedrooms and other nooks and crannies to heat, cold spots can be a real problem, regardless of what the literature says.

Remember: Unless the wood stove is fitted with baffles and an electric blower, which enable it to heat and circulate room air, you're relying mostly on radiant heat, which travels in straight lines and heats objects only. (There is, of course, some coincidental convective heating as room air circulates naturally across warm surfaces.) If the salesperson you're dealing with isn't much interested in the size of your rooms and their layout, you might want to try another shop.

Buying a stove with baffles and an electric blower can be a smart move, particularly if the stove's to be located in a relatively large room and/or you're interested in circulating warm air into other parts of the house. But beware of blowers that draw a lot of electricity or make a lot of noise.

For homeowners who already have a wood stove and find that its heat isn't reaching certain parts of their home, here are some ideas that may help.

- To get more heat into an upstairs room, you can put a transfer grille in the floor that's manually opened and closed as needed. If you have a cold room abutting a warm room on the same level, a transfer grille could be fitted into the wall that separates the two. These remedies work best with wood stoves that are designed to circulate heated air.
- Remember that doors act as dampers. Leaving them open or closed has a powerful effect on the way heat moves through your house. In some cases, where doors are very snug in their frames, it may be helpful to cut some off the bottom so that air can circulate into the room even when the door is closed.
- Tiny, quiet, low-amp electric fans mounted in doorways or inside transfer grilles can sometimes be effective in promoting the distribution of warm air.
- In rooms with high ceilings (10 feet or more), heat stratification may be a problem. A ceiling fan can help redistribute warm air back down into the room where it's needed.
- Re-insulating and/or weathertightening a cold room will help keep it warmer — maybe enough to solve the problem.
- As a last resort, you can address cold spots with some type of stand-alone electric or vented gas heater. (More on this later.)

Pellet Stoves

Pellet stoves, first introduced to residential markets in 1982, are quite different from and considerably more complex than wood stoves. Instead of using large chunks of natural wood, pellet stoves burn saw-

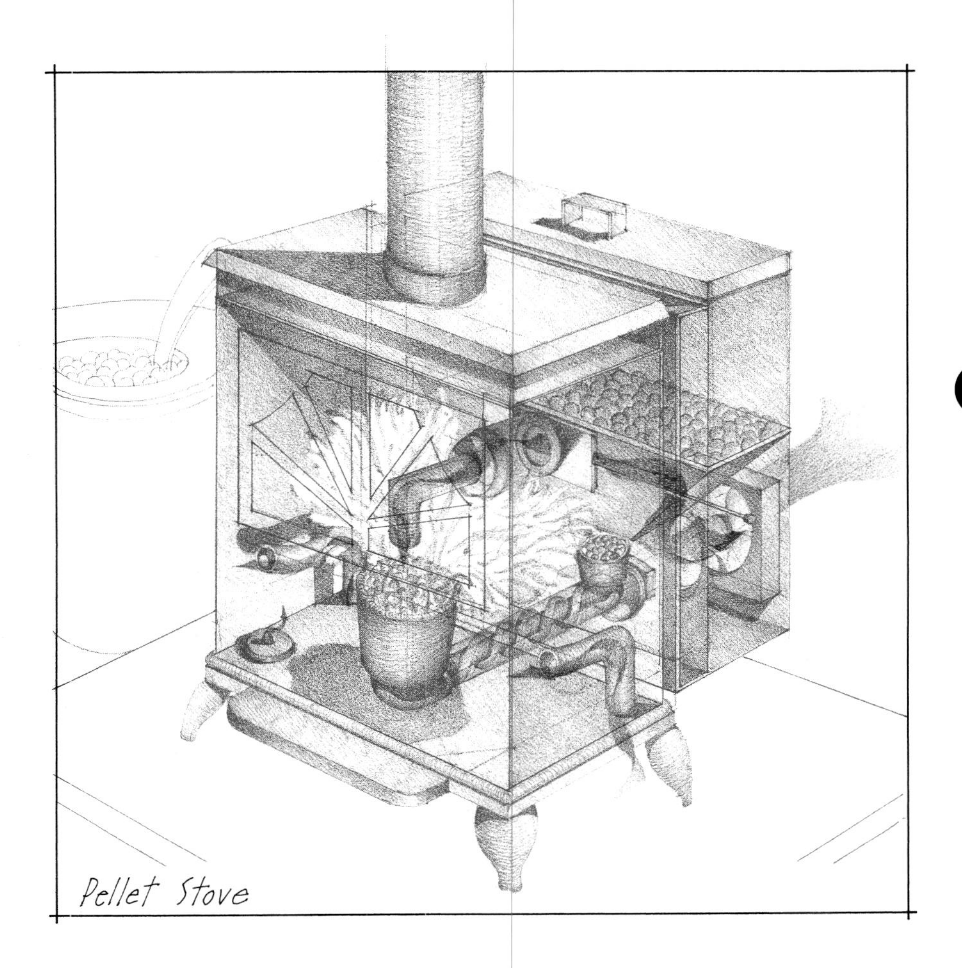

dust, recycled cardboard, and agricultural wastes that are formed into little pellets. (At first glance, they look like some kind of dry dog food.) The pellets are sold through stove shops and fuel distributors in 40- and 50-pound bags. The price per ton ranges from $130 to $225 depending on the ash content of the fuel and the distance from the pellet mill.

Freestanding models and fireplace inserts are available with 30,000, 40,000, and 50,000 BTU-per-hour ratings. Pellet furnaces and boilers offer BTU-per-hour ratings of 75,000 or more.

The most powerful advantage these stoves have is that they can burn up to two days at a steady, controllable rate without refueling. The pellets are loaded into a fuel bin that's built into the top or bottom of the stove. By means of an electric auger, or series of augers, the pellets are routed into a very hot and compact fire. The rate at which pellets and combustion air are fed into the firebox is set manually using built-in controls or adjusted automatically (as an option on some models) by means of a wall thermostat. Pellet stoves come equipped with baffles and an electric blower to promote convective heating.

To their credit, pellet stoves are cleaner than wood stoves

and can warm the soul (as well as the body) by providing a nice-looking fire through the viewing glass. Like natural wood, pellets are a renewable source of energy and are domestically produced. Unlike natural wood, pellets are clean and easy to handle — no splinters in the thumb, spiders in the woodbox, or pieces of broken bark littering up the floor.

On the down side, pellet stoves cost $1,400 to $3,500, which is quite a bit more than your typical wood stove. When overall efficiency and fuel prices are considered, they're also more expensive to operate.

One further disadvantage with pellet stoves — though not a big one in my view — is that they rely on electricity to operate, which can add $75 to $150 to your annual electric bill and leave your home without heat in the event of a power failure.

With any type of stove, whether catalytic, secondary combustion, or pellet, the safety of the installation is the *first* concern. The flooring and backing materials used for the hearth, the stove's clearance to surrounding objects, and the integrity of the chimney or vent pipe are key considerations.

If you decide to install a wood or pellet stove yourself, be sure to check local code requirements before you start and to follow the stove manufacturer's safety recommendations. After the installation is complete, ask a reputable dealer or someone from your local fire department to check it for you.

If you don't feel comfortable installing the stove yourself, the shop where you bought it will do it for you (for a fee) or refer you to a qualified subcontractor.

Wood and Gas Fireplaces

There are about 25 million traditional masonry fireplaces in the United States, some dating all the way back to the Revolutionary War. But whether the work was finished two hundred years ago or just last week, nothing is quite as beautiful to my eye as a finely crafted hearth made out of brick or stone. And nothing quite so romantic as a crackling wood fire rising behind a stately pair of andirons. Just the thought of it makes me want to pour myself a glass of Cabernet, stretch out on the rug, and spend some time reading one of the great classics . . . like the Massachusetts State Plumbing Code.

The problem with this picture isn't my choice of reading material, but the fact that a lot of beautiful fireplaces are actually disasters when it comes to heating. Sure, you can feel the radiant heat on your cheeks and hands when the fire is blazing — great! — but what that open hearth is doing most of the time is sucking heated

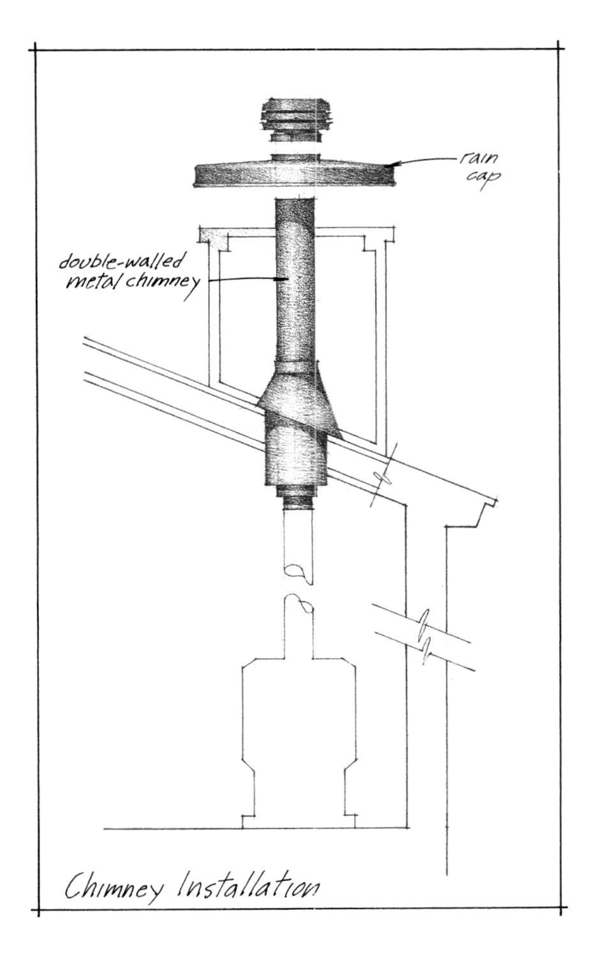

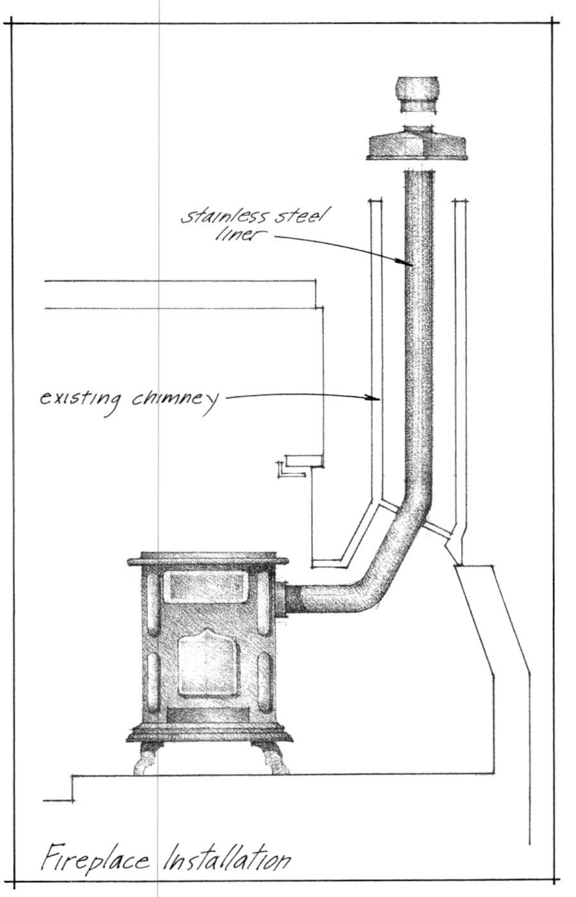

room air up through the chimney and out into the starry night. Tests show that traditional open-faced fireplaces have heating efficiencies ranging from 25 percent on the high side to *−10 percent* on the low!

In the 1970s and early '80s, when the United States suffered through oil embargoes and began to experience other energy price and supply problems, fireplaces fell out of vogue because they were perceived (correctly) as energy hogs. Not only did builders stop installing them in new houses, but a lot of fireplaces in existing houses were sealed up in the name of energy conservation. Others underwent flue modifications so that they could be fitted with a wood stove or fireplace insert.

Over the past few years the pendulum has swung back the other way. In a recent poll conducted by the National Association of Home Builders, homeowners voted fireplaces the most desirable special feature they could own in a home. This newfound popularity is borne out in recent construction statistics, which show that about six out of every ten new homes now come with a fireplace.

What the statistics *don't* show are the dramatic changes that have come about in the way fireplaces are being built and used.

First of all, masonry fireplaces are a dying breed. While I lament the decline of this proud old craft, it's hard for me to argue the bottom line. By the time you pay a mason for his hours and materials, and finish equipping your new fireplace with custom-made doors and other goodies, you can end up spending three or four times what you'd pay for a nice factory-built model.

Moreover, factory-built fireplaces are designed and tested as a system, which improves their safety and performance. Masonry has no such testing standards.

Flexibility is another plus. Factory-built fireplaces — or "zero-clearance" fireplaces, as they're sometimes called — can go into tight and unlikely places where masonry wouldn't be practical. Since the outer surface of the metal firebox is designed to stay cool, factory-built fireplaces can be installed within inches of wood and other combustible materials — hence the name "zero clearance."

And since factory-built units can work with a lightweight metallic flue instead of a masonry chimney, they usually don't need new foundation work or structural reinforcement to brace up the floor. These advantages save time and money during installation and make it possible to install a fireplace in places you've never dreamed possible — even in upstairs bedrooms and baths.

Another dramatic change in the world of fireplaces has been the introduction and quick acceptance of natural gas–fired models. As a matter of fact, about 10 percent of all new fireplaces run on natural gas or propane. They mimic wood fireplaces (with varying degrees of success) by spreading flame up through a stack of artificial logs, which are made out of cement, ceramic, or a ceramic-fiberglass mix.

The advantages of using natural gas or propane are quick to see. No wood to lug, ashes to sweep, or chimney to clean. With the flip of a switch you've got an instant, no-muss, no-fuss fire. If you're a real couch potato, you can even get a fireplace that clicks on and off with a remote control.

With the invention of direct venting, homeowners can put a zero-clearance gas fireplace just about anywhere they choose — even underneath a window! That's because the flue on direct-vent models runs straight back from the firebox through an exterior wall, eliminating the need for a vertical chimney. Direct vent units also score well when it comes to efficiency, since they're designed to use outside combustion air and come equipped with airtight doors. These same features make direct-venting fireplaces especially appropriate for tightly built homes — since they don't draw their combustion air from the house, you haven't got any depressurization or heat-loss problems to deal with. And the fireplace can't spill carbon monoxide and other noxious gases into the room.

Shopping for a Fireplace

The first thing you have to decide in choosing a fireplace is how efficient you want it to be as a heater. Manufacturers offer a wide range of wood and gas models with different performance characteristics. Some are designed almost purely for their looks and romantic value; others produce a lot of heat.

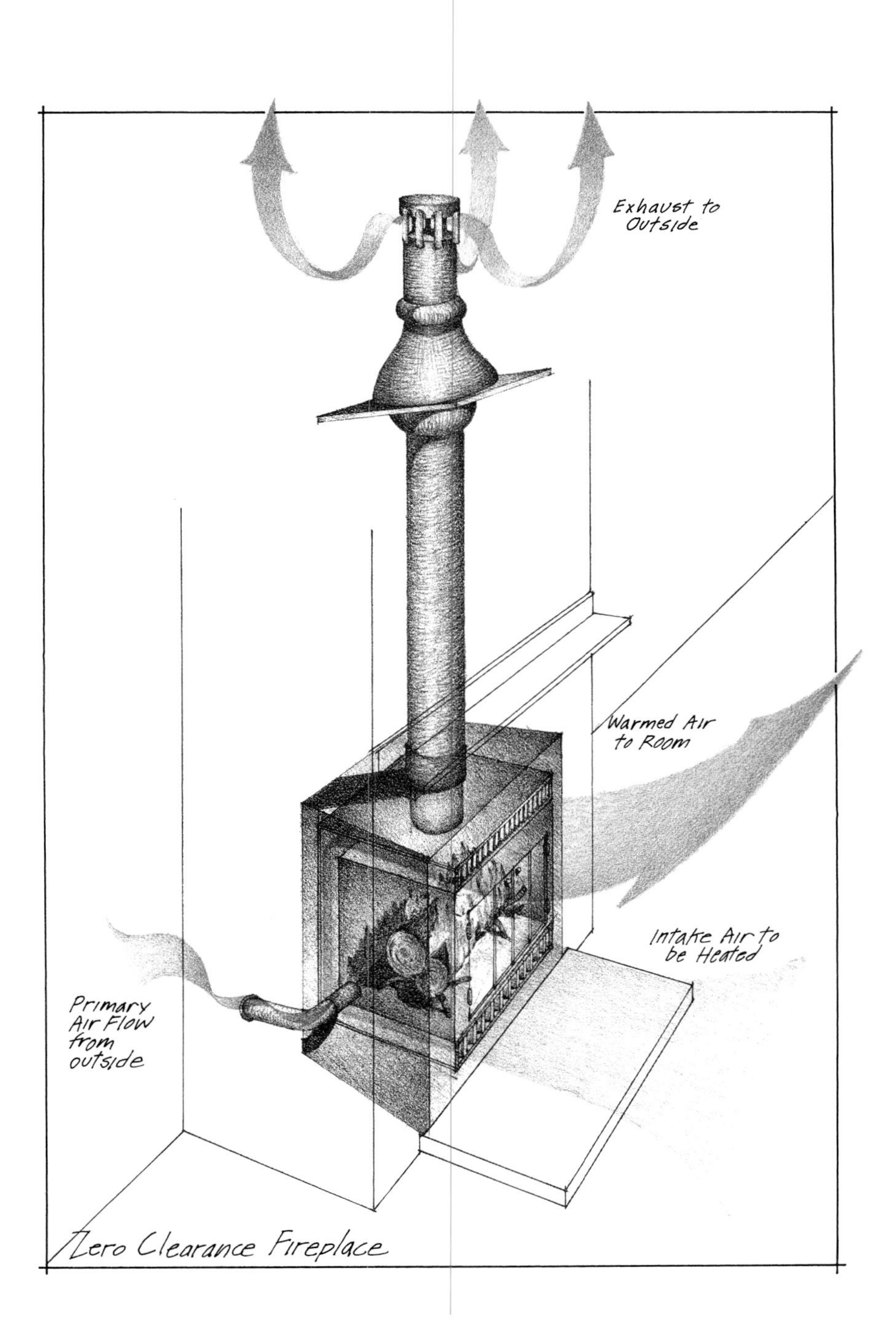

Zero Clearance Fireplace

Wood-burning fireplaces tend to have higher peak outputs than gas, up to about 70,000 BTUs per hour, but they have to be tended to maintain that rate. Gas models generate up to 30,000 BTUs, but burn continuously without tending.

Fireplaces that are designed with serious heating in mind will have most, if not all, of the following features:

- Sturdy glass doors with snug-fitting gaskets. I'd rate good doors as a "must" on *any* kind of fireplace because they keep heated room air from escaping up the flue when the fireplace isn't in use. In addition to their energy-saving role, doors provide a safeguard against flying sparks.
- Tight-fitting flue damper. A must for wood-burning fireplaces. (Top-venting gas fireplaces have their flue damper fixed permanently open as a safety precaution.)
- Outside combustion air. Fresh air is ducted to the firebox from outdoors so that the fire won't "steal" warm air from the house. Outside combustion air also insures that the fire will draw well, even in a tightly built home. If outside combustion air isn't used, the next best option is to have operable air intake vents on the fireplace that can be closed off when the fire isn't burning.
- Insulated firebox. Insulation placed between the walls of the firebox slows conductive heat losses through the metal and keeps cold air out.
- Built-in baffles for heat circulation. Some fireplaces draw room air through the baffles with an electric blower; others rely on natural convection. Tip: Just because a fireplace offers a "fan kit" option doesn't mean it will deliver a lot of heat. Check the stove's BTU-per-hour rating and its other energy features as well.
- Triple-wall flue. Helps prevent cold air from running down the chase and infiltrating the house.
- Masonry heat storage. A few very top-end, high-efficiency fireplaces use prefabricated masonry forms to provide heat storage.

Factory-built fireplaces (with flue) cost from $400 to $3,000, depending on the size, quality, heating efficiency, and finish. The low-end models — sometimes called "builder boxes" — are bare-bones units without doors. On the top end are modular heat-storing fireplaces, featuring ceramic glass doors, baffles, masonry heat storage, and (in some models) a fan/ductwork system for moving warm air into other rooms.

Good middle-range fireplaces equipped with tempered glass doors, outside combustion air, baffles, and a circulating fan, are available from $700 to $1800. The final *installed* cost of the fireplace will depend on whether you hire a contractor or do the work yourself, and the quality of the finishing materials you choose.

One of the first things you should look for when you go shopping — for safety's sake — is a certification label from Underwriters Labs (UL), the American Gas Association (AGA), or other accredited testing lab. Next, ask about the thickness of the steel and the overall weight of the fireplace. Sheer weight says a lot about how the unit was built.

Don't shop on price alone. Compare the efficiency of different fireplaces and take a few minutes to estimate what the annual operating costs are going to be. Ask some commonsense questions about the quality of the fireplace and the integrity of the dealer. Are the doors sturdy and tight-fitting? Is the blower noisy? How much electricity does it draw? What kind of warranty do you get? (The industry standard is 10 years.) How long have the dealer and the manufacturer he's representing been in business? Will they stand behind their warranty?

If you're going to buy a gas or propane fireplace, be sure to visit a showroom so that you can see the flame in operation. While every company claims to offer an "authentic-looking fire," some are more realistic than others.

Though I'm not by any means an interior decorator, I would suggest you buy a fireplace that's sized proportionately to the room you have in mind. A wide, protruding fireplace with a raised hearth might look great in a large room with high ceilings, while the very same installation would overwhelm a smaller room and look — well, silly. It might also produce more heat than you want. The most common size factory-built fireplace is 36 inches wide by 24 high, but there are also 24-, 30-, 42-, and 48-inch-wide units to choose from.

Some manufacturers offer multidimensional fireplaces that have glass on two, three, or even four sides. These models can be used to create beautiful partitions, which present the fire to two different rooms, or for an island fireplace with a 360-degree view. Generally, multidimensional and freestanding fireplaces work best in large rooms where they can act as a powerful — but not *over*powering — point of focus.

Most fireplace dealers offer various types of trim kits — with finishes in black, brass, chrome, or stainless steel — to match different decors. If one of these stock kits isn't your cup of tea, you can finish the hearth and surround with materials of your own choosing, such as pre-cut marble, tile, or a lightweight facade of brick or stone. As a crowning touch, you may want to add your own mantel. These are available new, including some very nice wood and marble reproductions. Or you can scrounge around the architectural salvage shops for an old beauty that's looking for a new home.

Installation

With a modest amount of skill and a few commonplace tools, you can install your own factory-built fireplace and save hundreds of dollars in contractor's fees. It's usually a two- to three-day job, depending on the complexity of the finishing work.

Most installation manuals include framing diagrams and other details to help the do-it-yourselfer along. (Take a good look at the manual *before* you buy, assessing both your own skills and the dealer's ability and willingness to help.) Here are some added tips:

- Check local fire and building codes before you start. Most jurisdictions will require a construction permit.
- Installing or extending gas lines for a gas fireplace is work for a *licensed plumber*. This is one job where safety *must* come first and do-it-yourselfers shouldn't tinker.
- If it's possible, locate the fireplace on an interior wall and run the flue up through the middle of the house. A warm flue draws better than a cold one and you avoid the costs of building an exterior chase. But remember: Metal flues are *not* zero-clearance. Joists and rafters that don't meet clearance specs (usually one or two inches) must be protected with a heat shield.
- The flexible duct that provides outside combustion air to the firebox can go through an outside wall or down into a vented crawl space. But *don't* link it to the attic or basement.
- If you think you might want to switch from wood to gas someday, go ahead and have a plumber lay in a gas line while construction is in progress. It'll make conversion a lot less expensive later on.
- Take special care to weatherseal the hole where the flue penetrates the wall or roof. The hole must be precisely cut, then properly sealed around the flue pipe with nonflammable insulation, flashing, and caulk designed for high-temperature applications.
- If you plan to build an exterior chase, ask your dealer what kind of insulation system is most appropriate to your climate. Insulation helps keep the flue warm and prevents cold air from leaking into the house.
- Don't neglect to install a rain cap on top of the flue.
- Ask a building inspector and/or someone from the fire department to inspect the job *before* you put in the insulation and drywall so that you can correct any mistakes in

the framing. Have a second inspection done when the job is completed.

Space Heaters

As the name implies, space heaters aren't intended for whole-house heating, but only as a supplemental source of heat. People generally use them in one of four ways, each of which involves certain trade-offs and compromises.

First and foremost, space heaters are used to warm up a cold spot that's not adequately heated by the central system. This is really a Band-Aid approach to solving a comfort problem — the better solution almost always is to invest in new weatherstripping and insulation for the problem area and/or to modify the central system so that it delivers more heat. The problem is, the better solution may take more time and money. It's a lot easier, in most people's view, to spring 50 or 60 bucks for a portable electric heater, plug the sucker in, and worry about the electric bills later.

Space heaters are also used to heat new additions, especially when the central heating system doesn't have enough extra capacity to handle the added space or when the cost of extending the central system would bust the budget.

On "This Old House" we've almost always chosen to modify and extend the central system rather than use a space heater, because we felt that the modified central system would provide greater comfort, economy, and safety over the long haul.

One instance where we did opt for a space heater was in the renovation of the old farmhouse in Arlington, Massachusetts — back in 1982 — when we used a vented kerosene heater for the detached garage. In that case, the garage was too far away from the house even to consider an extension.

A third reason that people buy space heaters is to save money when the central system isn't really needed. Imagine, for example, a single person who's only using two or three rooms in an eighteen-room house. A space heater — even an electric one — could heat the lived-in area for a fraction of what it would cost to fire up the central furnace and heat the whole house. The same money-saving principle might apply in the autumn or spring, when heating needs are often slight and sporadic.

Finally, there are instances, especially in the southernmost part of our country, where a space heater may be all the house needs, even in midwinter.

Shapes and Sizes

Space heaters come in 101 different varieties, including compact wall units, stand-alone models, portable electric heaters, and kick-space units designed to fit underneath a kitchen cabinet.

As with central systems, space heaters need to be properly sized for the job you have in mind. Buying too much or too little heating capacity for the room will only leave you dissatisfied and/or paying more than you should for fuel. Besides the dimensions of the space, some other key considerations in proper sizing are the room's weathertightness and insulation level, the amount of window area, the amount of exterior wall and its orientation to wintertime sun and wind, and of course, the climate. Lifestyle considerations, such as how you dress and how active you are, will also affect your comfort.

Heating capacity varies a lot from one space heater to the next, ranging from just 2050 BTUs an hour for a 600-watt electric heater to 65,000 BTUs an hour for a large gas-fired wall or floor furnace. On the high-capacity models, definitions begin to blur a little — a "space" heater that can put out 65,000 BTUs an hour could probably heat a small house.

Not only is the rated capacity of the space heater important, but so too is the way in which it delivers heat. Some types of space heaters, like quartz heaters and electric radiant panels, are designed mainly to heat objects. Their heat is transferred radiantly from the warm surface of the heater to the relatively cold surface of your skin.

Other types of space heaters deliver heat mainly through convection, that is, they're designed to heat room air as it passes across a hot surface. Some models use natural convection to move air across the heat exchanger (warm air rises, drawing relatively cooler air in behind it); other designs use an electric fan to push air through the heater.

A third type of space heater — let's call it a "hybrid" — is built to deliver both radiant *and* convective heat.

As a rule of thumb, I think radiant space heaters work best when the "heatee" (that's you) is going to be in close and fairly constant proximity to the heater — at a work station, say, or in a shop. Electric quartz heaters, for example, lose their effectiveness at about 15 feet. Also bear in mind that any part of your body that is shielded from the heater — by a desk, for example, or workbench — won't be warmed. And of course your body itself acts as a shield, so that the side away from the radiant source won't feel as warm as the side that's exposed.

Space heaters that rely on convection work best in rooms that can be sealed off from the rest of the house. That way you don't have a little heater trying to warm all the air in a large house.

The last option, and a good one in many circumstances, is to buy a space heater that uses both radiation and convection.

Trethewey's Tips: How Space Heaters Can Jeopardize Your Pipes

In addition to keeping you and your family warm in the winter, your central heating system has another important responsibility: *To keep the water pipes from freezing.*

But using a space heater can sometimes "fool" the central system into thinking it's doing the job when it's really not.

I've been called in on a lot of plumbing jobs where the pipes had frozen and burst, and the befuddled homeowners hadn't a clue as to how it had happened.

"We were plenty warm here in the living room and kitchen," they'd say.

The problem — you've probably guessed it already — was that they'd been using a space heater that either directly or indirectly warmed the area around the central system thermostat. In other words, the local, supplemental heat from the space heater had fooled the central system into thinking that the whole house was properly heated. Meanwhile a remote, seldom-used bathroom got cold enough for the pipes to freeze.

The moral: Use a space heater if you like, but don't forget that the rest of the house has water pipes, and they cost a pretty penny to replace.

Fuel and Safety

Out of the more than 58 million homes heated by portable or fixed space heaters in the United States, more than 25 million — or about 44 percent — use electric heaters. It's easy to see why electricity prevails over wood, natural gas, LP gas, and kerosene in this market. Electric space heaters don't require a flue, there are no fuel lines required or refueling to contend with, and there's less risk of fire or asphyxiation.

This last-mentioned point about safety is critical. A recent report issued by the National Fire Protection Association (NFPA) shows that a very high proportion of the fire-related deaths in American homes are caused by gas, kerosene, and electric space heaters. (Other leading causes, as I mentioned earlier, are wood stoves and fireplaces.)

Space heaters pose a greater risk than central heating systems, the NFPA says, because they provide so many more opportunities for error by the occupants — in installing them, maintaining them, fueling them, operating them, and especially in arranging household contents around them.

While electric space heaters aren't by any means risk-free, the NFPA's statistics show that homes heated by portable or fixed electric space heaters are less likely to have a fire-related death or injury than homes using space heaters fueled with natural gas, propane, or kerosene. The difference, of course, is that these fuels are combustible and produce poisonous gases as a combustion by-product.

In 1987 — the latest year for which I could find full data — household accidents involving natural gas, LP gas, and kerosene space heaters claimed 366 lives. Two hundred and seventy-two of those deaths occurred in fires; 94 more from carbon monoxide poisoning. Portable and fixed electric space heaters, by comparison, caused only 111 fire-related fatalities that year, even though electric space heaters are more common.

We don't know how many of those 366 deaths involved *unvented* gas and kerosene space heaters, but they clearly pose a much higher risk of asphyxiation.

Frankly, I'm mystified — or maybe "horrified" is closer to the mark — that unvented gas and kerosene space heaters are still legal for indoor use in some states. With no flue or chimney, these units spew their combustion gases right into the home. The manufacturers who make these things say they've reduced the risks by instructing homeowners to always leave a window cracked open in the room where the heater is operating and by prohibiting their use in bedrooms and bathrooms. (Too bad for you if you don't bother to read the instructions or follow the code!)

While it's true that unvented gas and kerosene heaters now come with oxygen depletion sensors that shut the unit off if the oxygen level falls below a certain threshold, it's also true that sensors can fail or be deactivated by a stupid but determined homeowner. And even if the sensor does work flawlessly, there may be serious health consequences that aren't immediately fatal.

As I understand it, the atmosphere around us is 21 percent oxygen, and when you start to burn that up in a closed space, the blood chemistry in humans begins to change. Tests have shown that when the concentration of oxygen in the air is reduced, people become distressed and less able to think clearly and work productively. In a worst-case scenario, with all the windows closed on a chilly night and the sensor failing, an ill-fated homeowner — cozied up to his unvented space heater — just gets drowsier and drowsier and drowsier. And the next thing he knows, he wakes up dead.

What I'm trying to say is that despite the fact that unvented gas, propane, and kerosene space heaters are still legal — and even approved by the American Gas Association and other certification bodies — I strongly discourage their use. The one exception, I suppose, would be for use in a *drafty* old barn or garage as a *temporary* source of *regional* heat.

While I'm still on the subject of safety, here are some other pointers to remember:

- Look for space heaters that are certified by the American Gas Association, Underwriters Labs, or other national testing laboratories.
- Shop for space heaters that have automatic shut-off features.

- Space heaters need space. I recommend a 36-inch clearance from any combustible material, including your body.
- Make sure that your home is well enough ventilated to preclude any danger of a backdraft. Better yet, get a space heater that draws its combustion air from outdoors.
- Follow the manufacturer's instructions on how to use and maintain the heater.
- Portable kerosene heaters are illegal in some areas. Check with local officials before you buy.
- Portable kerosene heaters need to be fueled in a well-ventilated area, free of heat sources, when the device is cool and only with the type of kerosene specified by the manufacturer. Never use gasoline or other substitute fuels.
- Don't use electric space heaters in the bathroom.
- Ask a building inspector or someone from the fire department to safety check your installation.
- Battery-powered or hard-wired smoke detectors should be standard gear on *every* floor of *every* house.

●

Water Heating

One of the crowning achievements of our society, at least from an engineering standpoint, is that we've turned hot water from a rare luxury into an everyday commodity. With the turn of a tap it's right there at our fingertips — in almost every home — to help us prepare the food, wash the dishes, and bathe the kids.

According to the Census Bureau, 79.8 million households in America enjoy hot water — that's 99 percent of us! (Back in 1940, by comparison, 30 percent of our homes didn't even have running water.)

Since the vast majority of us use conventional tank-type water heaters — either gas or electric — that's where we'll start our discussion.

Tank-Type Water Heaters

As you can see from the drawings, there are some close similarities between gas and electric tank-type water heaters, as well as some fundamental differences.

Gas water heaters have a burner situated beneath the tank, where natural gas (or propane) is mixed with air and burned. The ignition — usually a pilot light — is controlled by a thermostat that senses the water temperature inside the tank. When you shower or run the dishwasher, water flows out of the top of the tank, where it's hottest, and through the hot water pipes to the point of use.

Cold incoming water, flowing into the bottom of the tank through the dip tube, replaces what you use. When the thermostat senses that the water temperature in the tank has fallen below the preset limit, it fires the burner. Heat from the gas flame spreads through the metal bottom of the tank and into the water. The burner will keep on firing until the water temperature inside the tank meets the thermostat's setpoint.

The hot combustion gases released from the burner pass into a flue that (in most designs) runs up through the center of the water heater. A series of built-in baffles, or passageways, inside the flue extract additional heat from the hot gas before it leaves the water heater and is vented outside the house. Even so, the flue is a big source of lost heat and inefficiency.

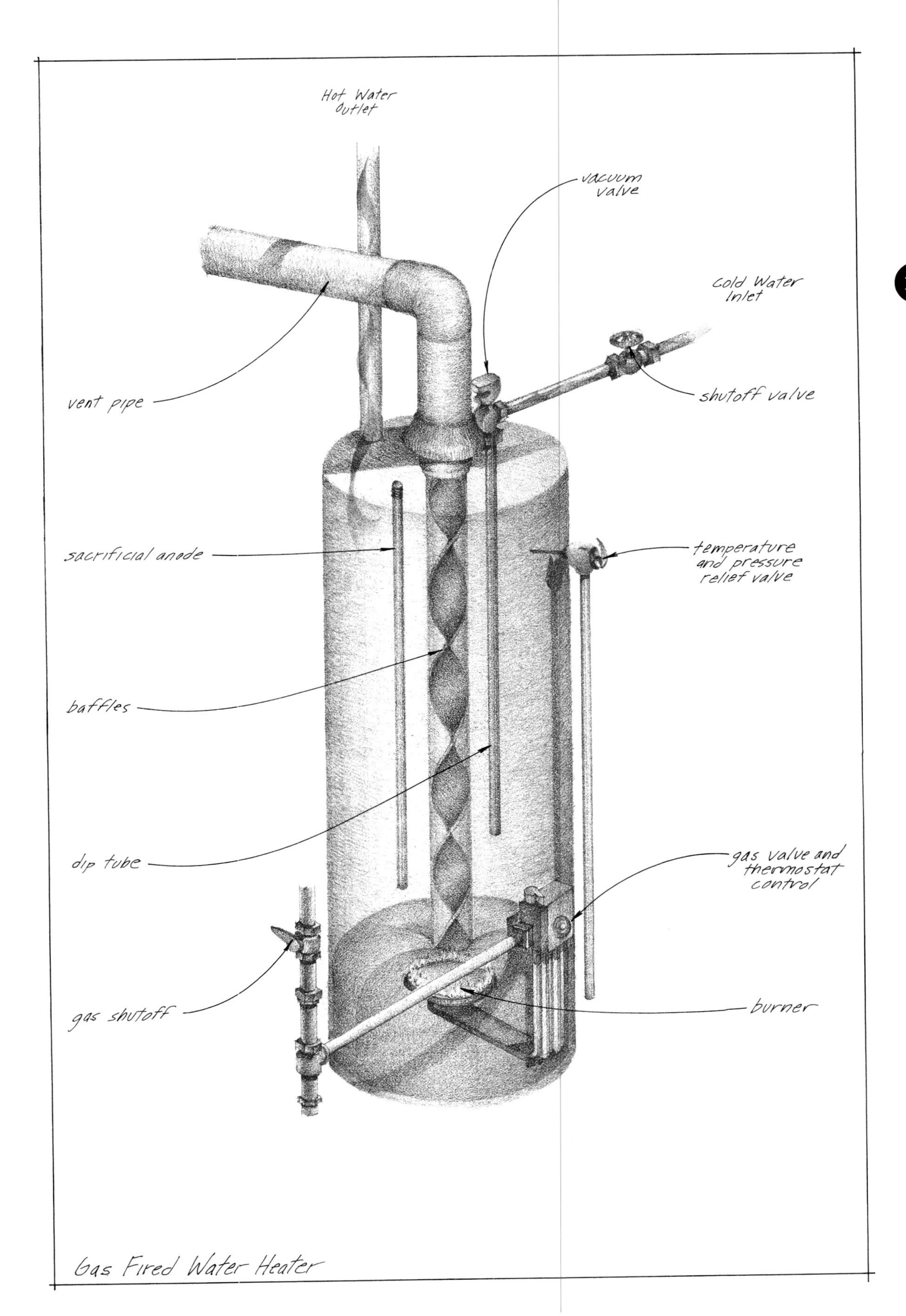
Hot Water Outlet
vacuum valve
Cold Water Inlet
vent pipe
shutoff valve
sacrificial anode
temperature and pressure relief valve
baffles
dip tube
gas valve and thermostat control
gas shutoff
burner
Gas Fired Water Heater

Electric water heaters, of course, have no burner or flue. They heat the water with one or two immersed elements. On the two-element designs, one is situated in the upper part of the tank and the other in the bottom, each with a separate control.

The lower heating element does most of the work, since it's the first one activated when you draw hot water out of the tank. Only when the tank has been practically emptied of hot water would the upper element be employed. In any event, the controls are designed so that the two elements never come on simultaneously.

The storage tanks on gas and electric water heaters are welded together out of rolled steel with an inner glazing of baked-on glass to protect the metal from corrosion. The outside of the tank is

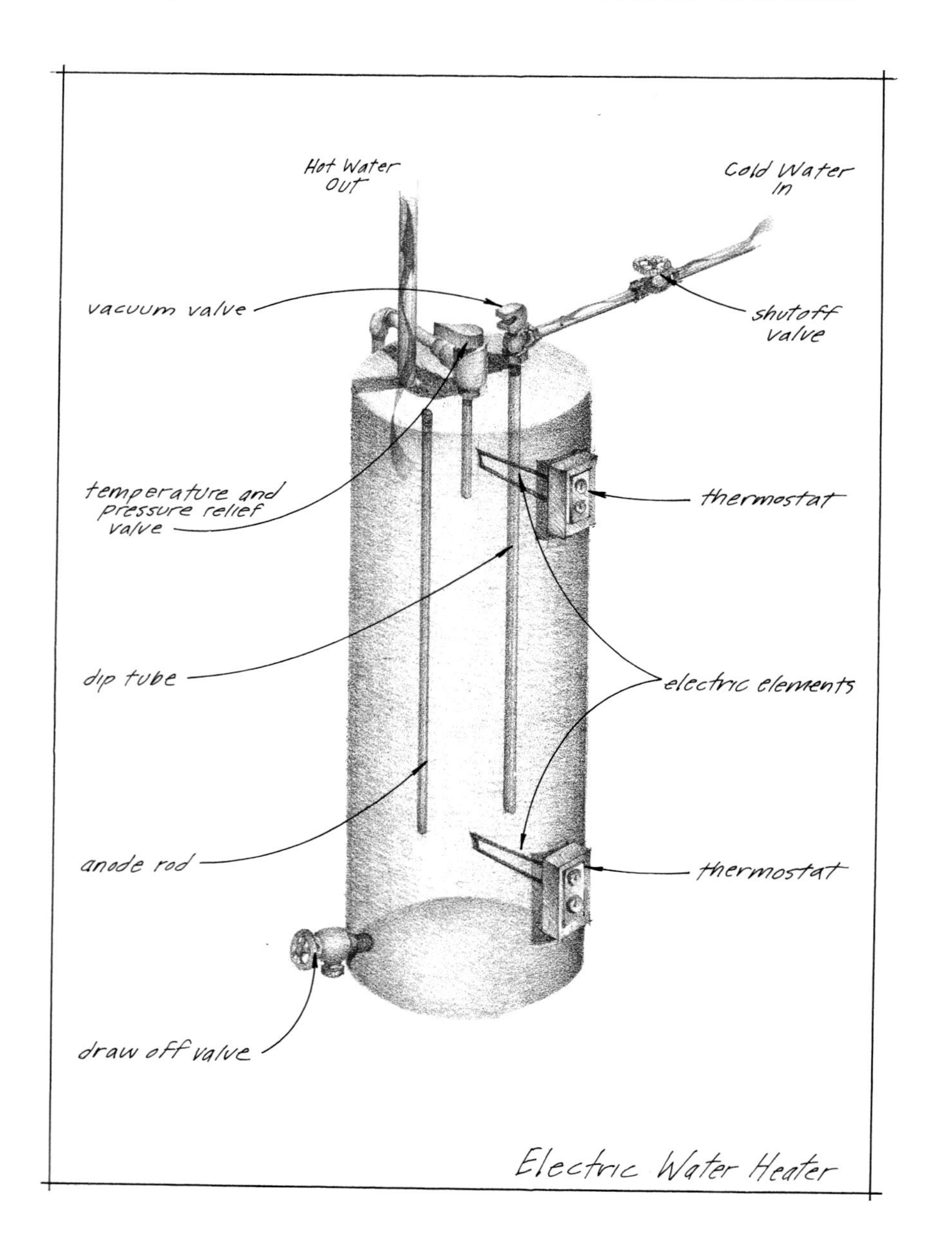

Electric Water Heater

factory-wrapped with a layer of fiberglass, mineral wool, or foam insulation, and then covered with a sheet metal jacket. Though rare and relatively expensive, some water heaters come with stainless steel or copper tanks, which are more resistant to corrosion. Steel tanks with an inner coating of plastic are yet another option, which I'll talk about later.

Gas water heaters come with 30-, 40-, 50-, 65-, and 75-gallon storage tanks. Electric models are available with 20-, 30-, 40-, 52-, 66-, and 80-gallon tanks. A family of three or four would typically use a 40-gallon gas water heater or an electric model with 52-gallon storage.

A critical piece of safety gear on all storage-type water heaters is the temperature and pressure valve, or "T&P valve" as we call it in the trade. Mounted on the top or side of the tank, the T&P valve is designed to prevent a serious accident should the water heater malfunction in a way that produces dangerously high temperatures and pressure. The valve is designed to pop open if the water temperature near the top of the tank hits 210°F (just below boiling) or if pressure reaches 150 pounds per square inch. In other words, the T&P's job is to blow hot water out of the tank and into a drain line before there's any chance of explosion.

A second safety device (required in some code jurisdictions) is the vacuum breaker, a brass valve that's mounted on the cold water inlet. As the name implies, the vacuum breaker ensures that if there's a sudden, large discharge of water (a broken pipe, say), the tank won't rupture as a result of the sudden vacuum.

Another component that's common to most tank-type water heaters is the sacrificial anode. As the name implies, its function is to "sacrifice" itself to the corrosive elements in the water so that the tank itself won't rust. The rate at which the anode dissolves depends on the water's mineral content, temperature, and hardness.

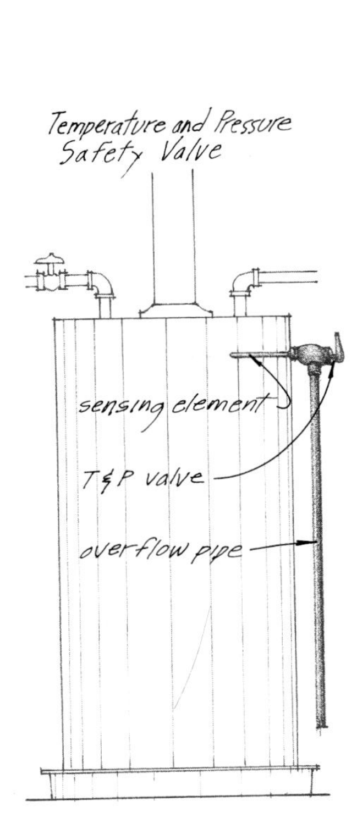

Anode rods are made out of magnesium, aluminum, or zinc. The rod on a typical 40-gallon water heater would measure about ¾ of an inch in diameter by 44 inches long, extending down through the top of the tank almost to the bottom. You might think of the sacrificial anode as your water heater's personal bodyguard — if the anode gets eaten away with rust and isn't replaced, the tank itself comes under attack.

A fourth fitting that comes standard on all tank-type water heaters is the drain valve, which is mounted near the bottom of the tank so that sediment can be periodically flushed out. Though most homeowners don't even know the valve is there — let alone how to use it — it's actually an important key to stretching your water heater's life and saving you money.

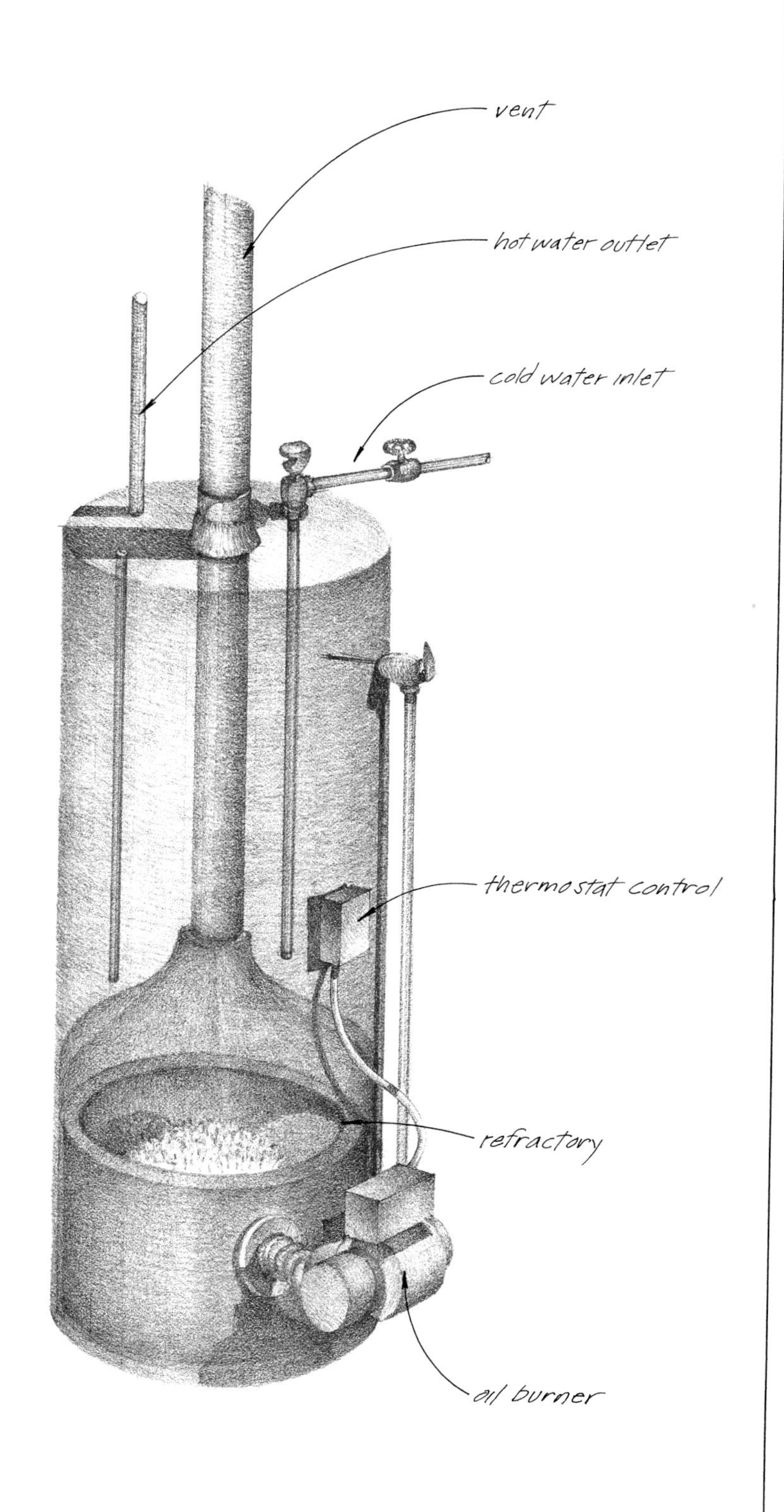

Oil Fired Water Heater

Before I get on with that, I'd like to say a few words about oil-fired water heaters. Though they can produce a gargantuan amount of hot water — more than any other type water heater — the tanks fail prematurely — sometimes in less than five years — because they're subjected to such intense heat. With that in mind, I don't advocate oil-fired water heaters for residential use. If you use fuel oil for your boiler, a better way to heat water for household use is with an indirect tank, which I'll discuss later.

Maintenance

The failure to properly maintain water heaters has turned them into one of the great triumphs (or is it *disasters?)* of our throw-away society.

Every five or ten years we put ourselves through this expensive and disruptive ritual of having to tear out the old heater and put in a new one. And it's not just the replacement costs I'm concerned about — I've seen beautifully finished basements and other rooms thoroughly trashed by leaking heaters.

More often than not, people panic at the sight of a leak and the thought of having to go without hot water for a day or two. That's when the let's-rush-out-and-buy-whatever-we-can-lay-our-hands-on-and-hastily-hire-a-plumber syndrome takes over, which sets the stage for the next failure.

Looking at the latest statistics from the Gas Appliance Manufacturers Association, we find that about 5.7 *million* Americans go through this nasty little ordeal every year. That's *not* counting the water heaters that are sold for new construction or into the industrial and commercial markets — just the replacements that we purchase for our homes.

Statistically, that means Americans are buying about 650 new water heaters an hour! (And presumedly having 650 *failed* water heaters carted off to the landfill or junkyard!) If the average installed cost of those new heaters is reckoned at $500 apiece, it means we're spending $325,000 an hour just to replace our water heaters. (Not to mention the added dollars laid out for disposing of the old ones.) Now $325,000 an hour may not be big money to the Pentagon or the Sultan of Brunei, but if you happen to be one of those 650 people who's paying the bill, it probably feels like *real* money.

My point here is that if a water heater is properly maintained, there's no reason that it can't last 20 or 30 or even 40 years. The problem is, almost no one does the maintenance. That's because most people don't understand that there are things they can do to extend the tank's life or they figure they're not going to live in the house long enough to reap the benefits. ("Hey, let the next guy take care of it. . . .") Or maybe they have so much money they don't care how often the water heater fails. (Unlikely, I think.)

Manufacturers and plumbing contractors could do a lot to educate homeowners in this regard, but they don't see as much profit in supplying parts and maintenance for old water heaters as there is in selling and installing 5.7 million new ones every year.

So here are some key points to remember if you'd like to double or triple the life of your water heater and avoid the flooded basement and panicky replacement ordeal that I described earlier.

The first step is to inspect the overall condition of the heater. An aged water heater that's already showing signs of rust around the pipes and fittings may not be worth the money and effort it takes to maintain it.

If you have an electric water heater, take the upper and lower hatch covers off the side and check for rust and leaks around the gaskets where the elements fit into the tank.

To gauge the condition of a gas water heater, open the combustion chamber hatch covers and peek inside with a flashlight. If you spot heavy rust or water marks, the tank is probably too far gone to worry about. (To save yourself headaches later on, it might be smart to go ahead and replace the old water heater *before* it starts to leak. This will give you time to shop for the best kind of tank, at a good price, without any pressure.)

If the tank is in good shape, and you're going to work on it yourself, make *absolutely* sure that you *turn the gas or electricity off before you start.* In some cases you'll need to let the water cool down or drain the tank before you begin. Think ahead. This is not kid's stuff. If you're careless or inept, you run the risk of being shocked or scalded, flooding the basement, or even causing an explosion. If you're not completely confident about doing the work, call in a pro and watch him. Ask questions. Take notes. Next time around you may be able to handle some or all of the maintenance work yourself.

Here's a checklist to help you along:

1. *Maintain the anode.* Inspect the sacrificial anode on your tank every three to five years and replace it if necessary. (If you have acidic water or use a water softener, the inspections need to be more frequent.) A good anode will protect the tank from rust, which is the most common cause of death among water heaters.

The anode typically has a ¾-inch hex-head fitting on top that's screwed down into the top of the water heater. You may have to use some muscle and a breaker bar to loosen it — then slide the anode up and out of the tank. On some water heaters the hex head may be recessed inside a hole in the metal jacket and covered by a

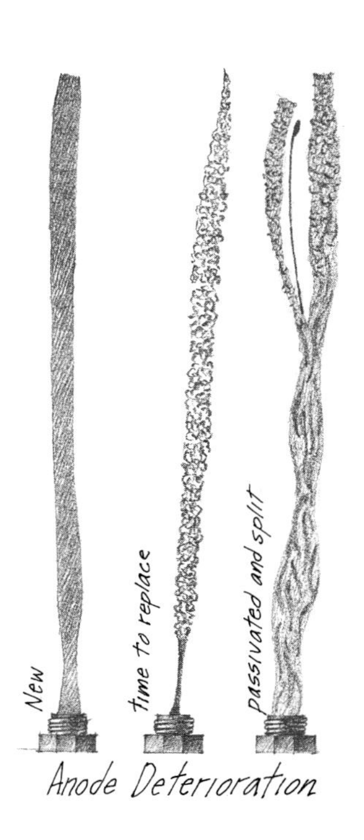

Anode Deterioration

Part Suppliers

If you have trouble finding the proper anode, curved dip tube, or other water heater parts from local outlets, try one of the suppliers listed below.

Elemental Enterprises
P.O. Box 928
Monterey, CA 93942
(408) 394-7077

Rheem Water Heating Division
One Bell Road
Montgomery, AL 36117
(800) 621-5614

Limebusters
2127 S. First Street
San Jose, CA 95112
(408) 293-1159

Perfection Corporation
222 Lake Street
Madison, OH 44057
(216) 428-1171

layer of fiberglass insulation and/or a plastic cap. You'll need to remove the cap and insulation so that you can get a socket wrench down on the head. On other types of water heaters the anode is built into the hot water outlet nipple, which has to be removed in order to reveal the anode.

Once you've exposed the anode, inspect it carefully. If it shows any bare wire or is "passivated" — that is, calcified to the point where scales flake off when the rod is slightly bent — it needs to be replaced. Some plumbing supply houses carry new anodes in stock (at $15 to $30 apiece); others won't have a clue what you're talking about.

Even if it involves writing or calling the manufacturer, be sure that you know where you can get a properly sized replacement anode *before* you start working on the tank. (The list above includes some anode and dip tube suppliers who can help if local suppliers don't have what you need.)

If your water heater is wedged into a tight space where there's not much overhead room to work, you may have to bend the new anode, insert it partway into the tank, and then straighten it out against the opening before you can screw it in. For really tight spaces, you may want to buy a flexible anode that's jointed end to end — like links of sausage — so that it can bend. As a very last resort, you can always drain the tank, break the plumbing loose, and move the water heater to a roomier spot.

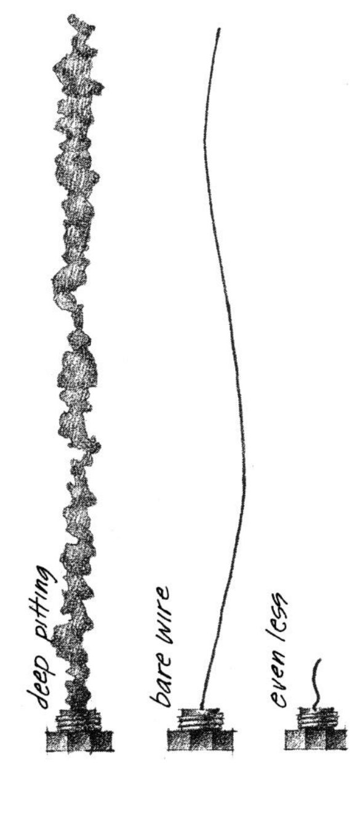

On some poorly designed water heaters the anode is actually sealed underneath the metal jacket. The only way to get at it is to cut away the metal jacket with a pair of tin snips, which is a pain in the whazoo. A better option is to replace the existing hot water nipple with a combination nipple/anode — that is, a pipe nipple with the new anode built right into it. What you're doing, in effect, is abandoning the old anode for a new one that's going to be easily accessible in the future.

As a rule of thumb, a water heater that's more than six years old will need a new anode. If softened water is being used, the anode could need replacement in just two years.

2. *Keep the tank free of sediment.* The second most important step in extending the life of your water heater is periodically to clean the sediment out of the tank. As a water heater ages, a sandy or gelatinous goo builds up in the bottom of the tank that's comprised of mineral deposits silting out of the water, little bits of metal and glass shed from the tank itself, and other miscellaneous sludge. Over time, the bottom of the tank can start to look like Mississippi River mud.

In electric models, the sediment eventually builds up to the point where it covers the lower element, causing it to overheat and fail. A sudden drop in your hot water supply may signal that the lower element has been drowned in sediment.

In gas water heaters, the sediment traps heat in the bottom of the tank, reducing the transfer of heat from the flame to the water and producing temperatures as high as 1000°F! That kind of heat is enough to stress the metal and cause the protective glass lining inside the bottom of the tank to dissolve, crack, or flake off, exposing the metal to rust. (Near the top of the tank, where the T&P valve is located, the water temperature may stay close to normal.) To make matters worse, steam bubbles pushing their way up through the sludge can make a hellacious racket, popping and cracking, as though someone were shooting marbles around inside the tank.

In any kind of water heater — gas, electric, or oil — sludge can become a happy breeding ground for certain types of bacteria, which can give a bad smell to the water. (If the odor persists *after* you've cleaned out the sediment, try changing the magnesium anode for one made out of zinc or aluminum.)

The traditional method of cleaning sediment out of a water heater is to turn off the fuel (but not the water supply), let the tank cool, and then flush a few gallons of water out the drain valve.

But Larry Weingarten, a nationally recognized expert on water heaters and co-author of *The Water Heater Workbook* (Elemental Enterprises, P.O. Box 928, Monterey, CA 93942), says that flushing does very little good in most water heaters, since it only removes the sediment that's built up close to the drain valve.

"Even after a thorough flushing, most of the sludge stays right where it is," he says. "Stuck around the edges of the tank."

To do the job right, Larry recommends that homeowners install a curved dip tube — a $5 item — to replace the straight dip tube that comes standard on most heaters. Water coming out of the curved tip creates a swirling effect inside the tank, he says, so that when you open the drain valve the sludge is effectively swept around and out.

The new tube is installed by unscrewing the existing cold water nipple and removing the old tube. If the old tube won't come out easily, you can coax it up by inserting a dowel or other cylindrical object, just to get a grip, and then walking the tube up out of the hole far enough to catch it with your fingers.

As shown in the drawing, the new dip tube needs to be oriented so that it directs inflowing water around the circumference of the tank, to sweep the sediment out. You can etch a little mark on the top lip of the dip tube fitting so that you know which way the tip is facing as you tighten it down. Be sure to wrap the threads on the nipple with six or seven turns of Teflon tape before you screw it into place — you'll get a much better seal.

If you really want to equip your water heater for effective and easy flushing, Larry suggests changing the cheap plastic drain valve on the tank with a new ¾-inch brass ball valve (about $8).

The plastic valves that come standard on most water heaters are notorious for clogging and dripping, and aren't very easily attached to a hose. A new ball valve will save you time and headaches by providing for a faster, safer, and more effective flush.

The new valve is best installed with the fuel and water supply shut off and the tank fully drained. The old plastic valve is removed by screwing it four to six turns counterclockwise while pulling out on it, then turning it clockwise while continuing to pull. Before you screw the new ball valve into place, be sure to wrap the threads on the nipple (and the adaptor, if one is used) with Teflon tape.

Curved Dip Tube

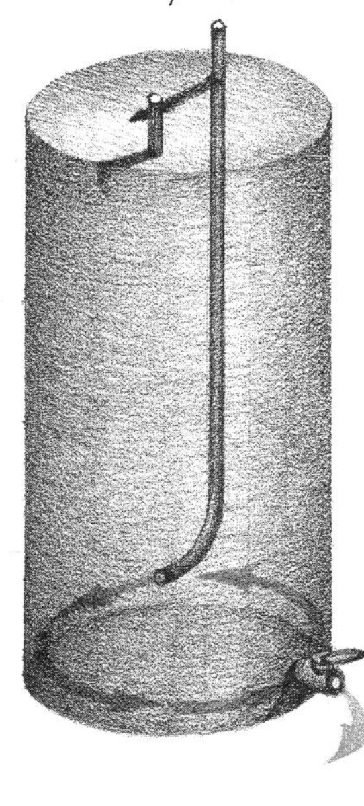

Once the curved dip tube and new valve are in place, it might be a good idea to treat the tank with a chemical descaler called Mag-Erad (available from distributors who handle A.O. Smith water heaters and related products). The application of this citric acid–based treatment (for gas water heaters only) is a bit involved, but it can remove really stubborn sediment that's baked on the tank.

Once you've got the tank really cleaned the first time, it's a snap — every six months or so — to connect a hose to the drain valve and let a few gallons of water flush through the tank until it runs clear. This usually means 3–5 minutes of running under pressure. With the curved dip tube, you can be sure that you're really doing some good.

An alternative method for cleaning an electric water heater is to remove the lower element from the tank and use a shop vacuum with a plastic extension to suck out the sediment. This has to be done with the *power off* and the *tank fully drained*. Since this method can get fairly laborious, repeated twice a year, it's probably less work over the long run to go ahead and change the dip tube and drain valve.

If the lower element on your electric water heater has failed, you can buy a new one for about $15 and replace it yourself. Again, remember that you're dealing with 240 volts, so make triple sure that the electricity is off and the tank fully drained before you change the element. If dealing with electricity and water leaves you spooked, you should let a plumber do it.

3. *Test the T&P valve*. For safety's sake, I recommend that you test the temperature and pressure valve on your water heater every time you flush it — that is, every six months. When you lift the steel lever on the valve, it should produce a healthy flow of water. If it fails to open or leaks after you've closed it, repeat the test a couple more times. If the valve won't respond properly, it needs to be replaced.

There's a 50-50 chance that in testing an old T&P valve — fused shut after so much time — you'll end up having to replace it because it won't reseat properly after you've forced it open. But that's a heck of a lot better than leaving the failed valve quietly in its place, which can lead to real disaster. If you do end up buying a new valve and test it regularly, it won't have a chance to fuse shut.

Putting in the Plug

When you consider that 15 to 40 percent of your home's total energy bill is paid out for hot water, you begin to understand how a little smart-sense conservation can save you big money over time. Furthermore, a few relatively small and painless things done to save energy and water in the home can have a very positive effect on our economy and environment.

The act of taking a shower, for example, can be deceptively simple. You peel down, turn on the faucet, adjust the temperature, and in you go.

But this ordinary little pleasure is actually underwritten by our country's ready access to fuel and water, neither of which is infinite in its supply or necessarily cheap. Contrary to what a lot of Americans seem to think, the right to a long hot shower has never been set forth in the U.S. Constitution or ordained by an act of God.

It is, however, a pleasure that we can sustain for ourselves and our children if we learn how to enjoy it with a little bit of common sense.

As it turns out, the three most cost-effective things you can do to save hot water are among the simplest and quickest jobs a do-it-yourselfer could ever imagine.

First of all, I recommend setting the thermostat on your water heater back into the 115–120°F range. Not only will this save you money on your fuel bill, but it will also reduce the chances of accidentally scalding yourself on water that's twenty or thirty degrees hotter than needed. Moreover, lower water temperature will prolong the life of your water heater, pipes, and fittings.

Some gas and electric utilities are willing to lower the water heater's setpoint free of charge. If your utility isn't willing, you can easily reset the tank yourself.

Since the temperature dial on most water heaters is inaccurate, the best way to make the adjustment is to open a hot water tap until it's reached its maximum temperature and then hold a candy or meat thermometer under the flow. If the reading is higher than 120°F, reduce the setting on the water heater. Gas water heaters are equipped with a single thermostat, located near the bottom of the tank. Electric models usually have a separate adjustment for the upper and lower elements, which need to be set in tandem.

After you've lowered the temperature, let the tank adjust before you take another reading. You may have to fiddle with the setting two or three times before you get the water temperature to stabilize in the 115°-120°F range. If you find that the new setting is a bit too hot or cold to suit you, you can always nudge it up or down a little.

The beauty in this change is that it costs you nothing — except a little time — but can make a big difference in your energy consumption. For one thing, your water heater doesn't have to work so hard to reach its setpoint. And since the tank is storing water at a lower temperature day in and day out, it loses less heat to the surrounding air. Engineers like to call this "reducing the tank's *standby losses*."

In some cases, lowering the hot water temperature may affect your dishwasher's performance. If you notice that dishes aren't coming out as clean as before, use the "pots and pans" or "heavy-duty" cycle when you wash. It's designed to engage a booster heater inside the dishwasher that raises the water temperature to 140°F or more for better cleaning. If you have an older-style dishwasher without a pots-and-pans cycle (booster heater), you may want to leave your water heater set at a higher temperature. (If you do, be sure to educate everyone in the family, especially youngsters, about the dangers of scalding.)

Recently, a few private companies and trade associations have started lobbying *against* lowering water temperatures, claiming that higher temperatures (above 131°F) are needed to kill the bacteria that causes Legionnaires' disease. But Dr. Richard Stanwick, a leading authority on the subject, says that the real health risk is being scalded.

"What you're talking about here is the very small and unsubstantiated risk of contracting Legionnaires' disease in the home compared to the very large and substantiated risk of being scalded," says Dr. Stanwick. "The bacteria that causes Legionnaires' disease (*Legionella pneumophila*) is everywhere around us and would certainly not be eliminated from the home environment simply by raising the water temperature on the hot side of the plumbing. It makes good sense, both for safety's sake and energy conservation, to go ahead and lower the tank setting."

Dr. Stanwick's view is echoed by the Consumer Products Safety Commission and the American Academy of Pediatrics.

A second good way to conserve hot water with no sacrifice in comfort is to equip your baths with low-flow shower heads and put faucet aerators on the lavatories. Compared to old-fashioned shower heads, which can gobble up to 10 gallons of water a minute, low-flow heads use only 2.5 gallons or less.

It's a quick and simple matter to check the flow rate on your current shower fixture. First, open up the top of an empty half-gallon milk carton so the entire top forms a square. With the shower running at a fairly forceful rate, hold the carton underneath the head and see how long it takes to fill. If it fills in less than ten seconds, your shower head is a bona fide water guzzler.

Low-flow shower heads and faucet aerators both rely on the same working principle: By drawing air into the water, they reduce the flow rate and give the water a softer, bubblier feel.

Low-flow shower heads cost $10–$25 apiece and will pay for themselves in saved water and energy in thirty to sixty days. Be sure to get one with a stop-flow button on it, which lets you stop the water while you lather up or shave without having to readjust the temperature when you turn the water back on. (Designs with adjustable microheads and other fancy but unnecessary stuff can cost up to $60.) Taking the old shower head off and screwing the new one on is about a twenty-minute job, requiring a pipe wrench and a roll of Teflon tape to wrap the threads.

A faucet aerator for your bathroom lavatory will typically cost $2–$3. Aerators for kitchen sinks, equipped with an on-off switch, run about $8. Installation is simply a matter of unscrewing the old faucet head and screwing in the new one. Even if you stop to ponder the macroeconomic consequences of your action, the job shouldn't take more than ten minutes.

There are other ways you can modify a shower head or faucet, of course — with flow restrictors or regulators — but the results won't be as good. The great thing about low-flow shower heads and faucet aerators is that they're engineered to produce a stream of water that's every bit as comfortable as the old-fashioned fixtures. Maybe more so. And they do it using a lot less fuel and water.

A recent study conducted by The Earth Works Group found that a family of four, in which each member took a five-minute shower daily, saved 27 cents a day in water and 51 cents a day in electricity just by switching to a low-flow shower head. Your own savings might be more or less, of course, but when you multiply them out by the months and years to come, they really start to add up.

To illustrate the effect this kind of *little* change can have on our nation's environment and natural resource base, consider the following: A family of four switching to a low-flow shower head could save 14,000 gallons of water a year. If 10,000 similar families were to do the same, the United States would save 140 million gallons each year. If 100,000 families made the switch, the total annual savings would soar to 1.4 *billion* gallons. Consider also the millions of dollars that could be saved in water purification and pumping expenses, which would benefit us all in the form of lower water bills and taxes!

On the air-pollution side of the equation, installing a low-flow shower head in your home will reduce the amount of carbon dioxide released into the atmosphere by 80 pounds a year if you have a gas-fired water heater and by 300 pounds a year if it's electric. The less carbon dioxide we pump into the air, the less we contribute to the so-called Greenhouse Effect, which may be causing global warming.

Insulating Jacket

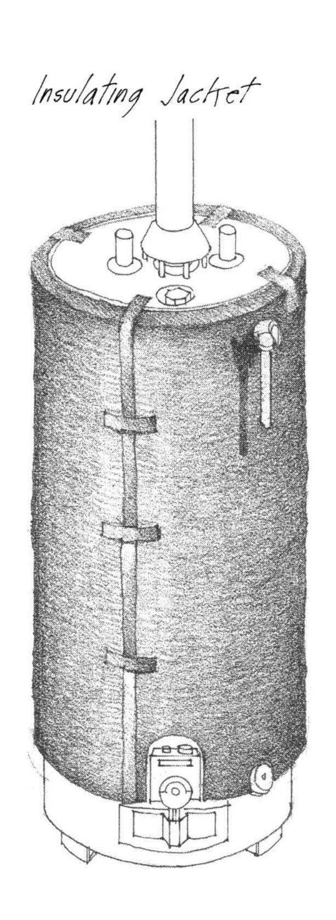

If the plumbing in your bathroom was properly done, installing a low-flow shower head shouldn't present any risk of scalding. Plumbing codes call for the installation of a thermostatic mixing valve, pressure-balancing valve, or an antiscald valve in the shower, any of which will prevent accidental scalding. However, if none of these devices is in place and the water supply to the bathroom is undersized (½ inch instead of ¾), and you've left the water heater set at 130°F or 140°F, installing a low-flow shower head is asking for trouble. Scalding could occur when there's a sudden drop in the line presure — say, a toilet flushing — that lets hot water force its way up the shower's water supply pipe, leaving only hot water coming out.

If you suspect your bathroom plumbing may be deficient, let the shower run over your hand, adjusted to the temperature you

Trethewey's Tips: 9 Wonderful Ways to Cut Your Water-Heating Costs

1. Lower the thermostat on your water heater so that it produces water in the 115° to 120°F range.
2. Wrap your water heater and pipes with insulation.
3. Equip your baths with low-flow shower heads. There's no comfort or convenience lost in the change — just lower water and fuel bills.
4. Ditto for low-flow faucet aerators.
5. Whenever you can, use cold water for the family wash and other cleaning chores. Many modern laundry detergents and household cleaners are made for cold water use.
6. Run your dishwasher to wash a full load of dishes whenever possible, using short cycles for everything but the dirtiest dishes, and letting the dishes air dry whenever you can.
7. Install heat traps on the hot and cold water lines coming out of your water heater. These U-shaped detours stop hot water from rising in the pipes when no one is using water. You can buy standard valves for the job (many new water heaters come with these already plumbed in) or make heat traps from long copper flex connectors bent into an inverted "U" shape. Either way, the fixtures will pay for themselves in less than a year and deliver savings happily ever after.
8. Have an electrician install a time clock on your water heater (about $30), which lets you schedule the heater's "on" and "off" times. Time clocks are most cost-effective on electric water heaters that can be programmed to benefit from special "off-peak" utility rates.
9. Be sure to turn the water heater off when you leave for the weekend or go away on vacation.

would normally use, while someone flushes the toilet. If there's a spike in the temperature with the old shower head in place, the risks of scalding would be increased with a low-flow shower head. To correct the problem, have a plumber install an antiscald valve on the shower or reset your water heated to the 115°F to 120°F range that I suggested earlier.

A third great way to cut your water heating costs is to put an insulating jacket around the water heater. Here again, the materials are inexpensive, the labor light, and the savings quite substantial, because you're cutting the heater's "standby" heat losses into the surrounding space.

Many electric and gas utilities will give you a ready-made water heater wrap for free. But even if you have to spend $20 or $30 to buy it or make it yourself, the investment will pay you back many times over.

If you buy a kit, opt for one with an insulation rating of at least R-11 and closely follow the instructions on how to install it.

A homemade wrap can be fashioned out of 6-inch-thick aluminum-faced fiberglass, which is taped snugly into place with the backing faced out.

With any type of wrap, it's absolutely critical that the insulation does not interfere with the safe and proper operation of the water heater. On gas- and oil-fired models, the insulation must in no way restrict the flow of air into the combustion chamber at the

Life-Cycle Costs of Various Water Heaters

Type	Efficiency	Approximate Installed Cost	Yearly Energy Cost[1]	Life (Years)	Cost over 13 Years[2]
Conventional gas storage	55%	$ 425	$190	13	$2,895
High-efficiency gas storage	60%	$ 500	$174	13	$2,762
Oil-fired free-standing	55%	$1,100	$228	8	$4,487
Conventional electric storage	90%	$ 425	$454	13	$6,327
High-efficiency electric storage	95%	$ 500	$430	13	$6,090
Demand gas	70%	$ 650	$160	20	$2,503
Demand electric (2 units)	100%	$ 600	$404	20	$5,642
Electric heat pump	200%	$1,200	$204	13	$3,852
Indirect tank with efficient gas or oil boiler	75%	$ 700	$148	30	$2,230
Solar with electric backup		$3,000	$144	20	$3,822

1. Energy costs based on the hot water needs of a typical family of four and energy costs of 8¢ a kilowatt hour for electricity, 60¢ a therm for gas, and $1.00 per gallon for fuel oil.
2. Future operating costs are neither discounted nor adjusted for inflation.
SOURCE: *Consumer Guide to Home Energy Savings*, American Council for an Energy-Efficient Economy.

bottom of the tank. Make sure that the insulation wrap is slightly compressed and snugly taped into place so that it doesn't sag down around the combustion chamber. And check it occasionally to make sure it hasn't slipped.

Likewise, the flue coming out of the top of the tank needs to have 3 or 4 inches of clearance so that the insulation won't interfere with proper venting. And be sure to cut the insulation back away from the controls, anode head, pipes, and — most important — the temperature and pressure relief valve.

Electric water heaters can be more completely wrapped top and bottom, so long as the controls, anode head, pipes, and the temperature and pressure relief valve are accessible and unobstructed. If you don't feel confident about installing the wrap, ask your local utility or state energy office for help.

Any water heater that meets the 1990 standards of the American Society of Heating, Refrigeration, and Air Conditioning Engineers (ASHRAE-90) won't need an additional insulation wrap, since it already has adequate insulation built in at the factory.

How to Size a New Water Heater

The best way to size a new water heater is to match it to your home's peak demand. Using the chart below, figure out how many gallons of hot water your family uses during the peak hour of usage on a busy day. After you've estimated the gallons used during that peak hour, all you have to do is select a water heater with a First Hour rating to match it. The First Hour rating takes into account both the tank size and the rate at which cold water is heated, so it's a much better indicator of capacity than the tank size alone.

Let's say, for example, that your peak hour includes four six-minute showers (as the family gets ready for work and school) and a hot water shave for Dad. Assuming that you have a low-flow shower head, we'll estimate usage for the showers at 72 gallons (4x6x3), plus two gallons more for the shave.

Since the total peak-hour usage comes to 74 gallons, you'd want to select a water heater with a First Hour capacity right around 74. (Note that without the low-flow shower head hot water usage could be as high as 170 gallons during that peak hour (4x6x7+2), outstripping the First Hour capacity of any water heater made. The last two people to hit the shower, in that case, would have to like it on the chilly side.)

First Hour ratings for different-size electric water heaters run from 16 to 94 gallons. For gas, the range is 52 to 86. Oil-fired water heaters have First Hour ratings above 110 gallons.

The Gas Appliance Manufacturers' Association publishes a complete directory of gas water heaters including First Hour ratings, energy efficiency, and estimated energy bills. (Send $5 to the Gas Appliance Manufacturers' Association, 1901 N. Fort Meyer Drive, Arlington, VA 22209. Ask for the "Directory of Certified Water Heater Efficiency Ratings.")

Where the Hot Water Goes

Use	Amount
Automatic clothes washer (warm wash/cold rinse)	10–12 gallons per load
Automatic clothes washer (cold wash/cold rinse)	0
Dishwasher	14 gallons per load
Hand dishwashing	4 gallons[1]
Tub bath	20 gallons
Shower	4–7 gallons a minute
Shower with low-flow head	1.5–3 gallons a minute
Bathing an infant	2 gallons
Shaving	2 gallons
Shampooing	4 gallons
Washing hands and face	4 gallons
Food preparation	5 gallons

1. Provided you fill the wash and rinse basins instead of letting the water run. Otherwise, you can end up using substantially *more* water than an automatic dishwasher.

I do recommend insulating your home's water pipes — wherever they're accessible — with foam pipe sleeves, which come in 3- to 6-foot lengths and can be clipped to fit with a pair of scissors. Hot water pipes are insulated to prevent heat loss between the tank and the final point of use. Cold water pipes are insulated to prevent condensation and dripping.

In the sidebar on page 184, I've outlined some other ways you can conserve water and fuel, and still have all the hot water you need.

Buying a New Water Heater

Old water heaters usually give up the ghost by springing a leak, which can range from a modest trickle to a dramatic flood.

If that happens, resist the urge to panic. Calmly turn off the water feed valve to the tank and then shut off the gas, electricity, or fuel oil supply. Attach a suitable length of hose to the drain valve and drain the remaining water out of the tank. (You may have to open a faucet to break the vacuum.)

After you've mopped up the water, try to *relax*. There's no need to rush out and buy a water heater on the spur of the moment. You and your family can survive a day without hot water, which will give you time to think things through and buy smart.

Here are some of the most important considerations in choosing a new water heater.

- First and foremost, consider the operating costs of a new water heater as well as its sticker price. If you simply buy the cheapest water heater around, ignoring the operating costs, you could end up with a real anti-bargain on your hands. As the chart on page 185 illustrates, the "life-cycle" cost of different types of water heaters can vary by as much as $4,000!
- Use the water heater's First Hour rating as your guide to proper sizing. The sidebar on page 186 explains how to do this.
- Consider the model's Energy Factor (EF) as an indication of overall efficiency. The EF will be listed, along with the First Hour rating, on the water heater's EnergyGuide sticker, which appears on all new tank-type water heaters. The chart on page 189 provides a high-low range of EFs for gas, electric, and oil water heaters, corresponding to different tank sizes.

The EF is *not* a good yardstick, however, when comparing water heaters that use different fuels. Even though the EF on a good-quality electric water heater might be 97, a gas-fired model with an EF of 63 would probably be more economical.

Why? Because electricity can cost three or four times as much as natural gas when compared on a BTU-to-BTU basis.

In other words, an electric water heater may have a higher EF and cost you a little less up front, but a gas water heater will only cost you half to two-thirds as much to operate. A typical family of four — paying 8 cents a kilowatt hour for electricity — will spend more than $6,000 for hot water over the estimated thirteen-year life of an electric water heater. The same family, using a gas water heater and paying 60 cents a therm for gas, would spend less than $2,800 — a savings of more than $3,200!

The recent introduction of direct-vent gas water heaters, which use a horizontal flue, has made it possible for homeowners to

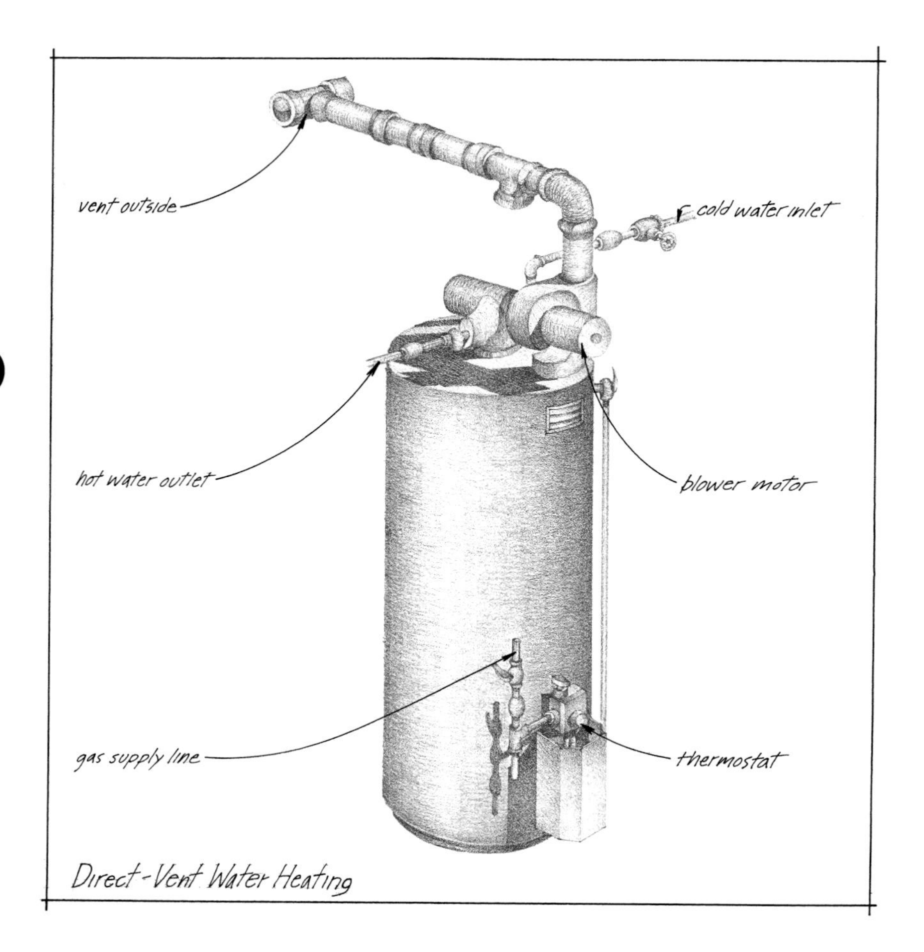

Direct-Vent Water Heating

switch to gas or propane without building a masonry chimney or running a flue pipe up through the roof. As shown in the drawing on page 188, the vent can be simply routed through the nearest exterior wall.

Some direct-vent water heaters have sealed combustion, which means they draw their combustion air from *outside* the house. I like this idea a lot. Not only does it make the water heater more efficient, it also rules out any chance of dangerous backdrafts. As I mentioned before in discussing furnaces and boilers, I think sealed combustion will become the standard on all combustion appliances within a few years.

As a rule of thumb, the more hot water you use in your home, the more you ought to be concerned about efficiency. If you have three baths, a hot tub, and four teenagers, you *need* efficiency with a capital E! Since most HVAC contractors and homebuilders use the Q&D method (Quick & Dirty) for selecting water heaters and other appliances, you may have to take the initiative yourself.

Prices and Features

Water heaters cost from $150 to $600, depending on the type and where you buy it. (The more middlemen you go through, the higher

Water Heater Efficiency (Energy Factors)

Tank Size	Gas		Electric		Oil	
	Low[1]	High	Low[1]	High	Low[1]	High
30 gallons	.56	.65	.91	.98	.53	.63
40 gallons	.54	.74	.90	.97	.53	[2]
50 gallons	.53	.71	.88	.96	.50	[2]
60 gallons	.51	[2]	.87	.96	.48	[2]

1. The minimum Energy Factor (efficiency) allowed under the National Appliance Energy Conservation Act.
2. No models available in this tank size.
SOURCE: *Consumer Guide to Home Energy Savings*, American Council for an Energy-Efficient Economy

the price.) When you add in the labor — usually two to four hours — total installed costs range from $250 to $800.

A run-of-the-mill water heater will typically come with R-6 or R-8 insulation and a 5-year warranty. Top-quality models will have R-16 insulation and an 8- or 10-year warranty.

State Industries makes a line of electric water heaters that are guaranteed against leaks for as long as you live in your house. The tanks have an inner lining made of Duron, a rugged plastic that won't corrode. Duron isn't available yet on State's gas-fired water heaters, but the company says that it's working on it.

Another nice feature that comes standard on some of State's top-line water heaters — called the "Turbo" line — is a curved dip tube that helps keep the tank free of sediment.

Rheem Manufacturing offers both gas and electric models with lifetime guarantees. The Performer series (also sold as the Kenmore Survivor) features storage tanks made out of fiberglass-reinforced polybutylene, which is impervious to corrosion. As shown in the drawing, the burner and flue on the gas models are located *outside* the storage tank, with a small pump circulating water between the tank and the burner. Without a central flue running up through the middle of the tank, standby losses are cut to a minimum, which is one of the reasons that the Performer is among the most efficient water heaters in the industry.

If you haven't got the time, skill, or inclination to properly maintain a conventional glass-lined water heater and expect to live in your current house for a good while, it would probably be worth your while to pay extra for a plastic tank with a lifetime guarantee.

No matter what brand you end up selecting, make sure that the anode rod is easily accessible at the top of the tank. (The plastic tanks I mentioned above don't have anodes, since there's no real threat of corrosion.) Also be sure that your new water heater meets ASHRAE-90 insulation standards.

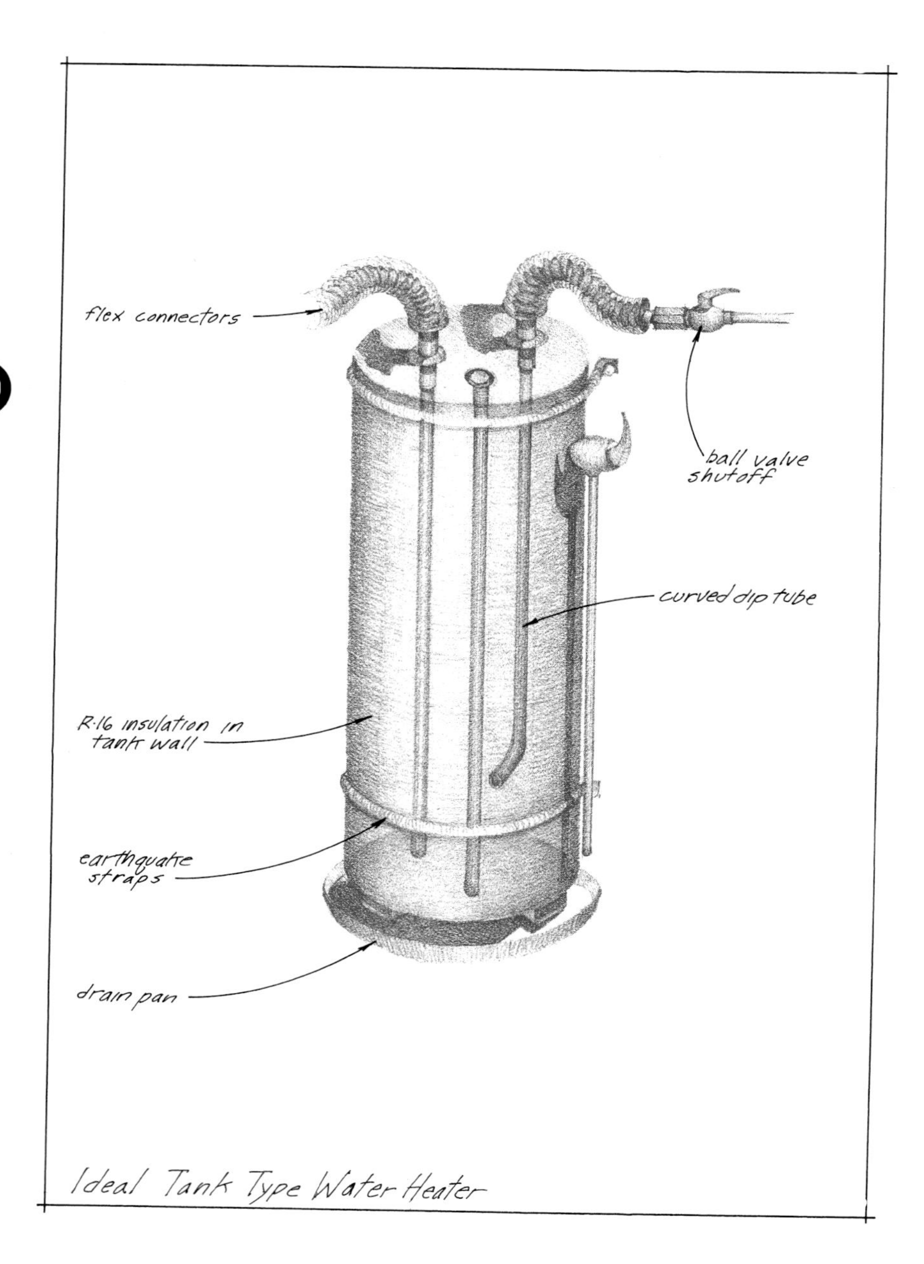

Ideal Tank Type Water Heater

Choose a competent installer with a good reputation, who's willing and able to make the following modifications to the water heater during installation.

- Replace the straight dip tube with a curved one. (This wouldn't be necessary with State's Turbo heater and some Kenmore models.)
- Remove the cheap plastic drain valve that comes on the tank and put in a ¾-inch brass ball valve.
- Install heat traps on the hot and cold water lines coming into the tank. Ball check valves can be used for this, but can make a chattering noise. A quieter solution is to use

flex connectors bent into an upside-down "U" shape. (See drawing on page 190.)

- Dielectric unions and plastic-lined steel nipples should be used so that the steel tank doesn't come in direct contact with copper and brass fittings. This is to avoid electrolytic corrosion, which occurs when two dissimilar metals are in contact.
- Place a drain pan, fitted with an appropriate drain line, underneath the water heater.
- The water shutoff valve needs to be located where it's readily accessible.

If you buy a gas- or propane-fired water heater that doesn't have sealed combustion, you and your installer should make absolutely sure that it's going to have plenty of combustion air. Gas water heaters should never be installed in closets or other tightly enclosed areas without proper ventilation. Remember that furnaces, fireplaces, clothes dryers, range vent hoods, attic fans, and bathroom exhaust fans all "compete" for indoor air — especially in tight houses — and can "conspire" to create a dangerous depressurization around the water heater. If the water heater doesn't have plenty of combustion air and a properly installed flue, it could backdraft deadly carbon monoxide into your home.

Where to Put the Thing

Unless you're building a new house, there's usually not a lot of choice in positioning a new water heater. It generally takes the place where the old one expired. But to the maximum extent possible, I urge you to modify the installation for easy access. This is particularly important if you're going to take my earlier advice and do regular maintenance on the heater.

It is important that not only water heaters, but any type of HVAC gear, be installed for the easiest possible access. By thinking ahead, you can save yourself and others an awful lot of hard work and money.

One of the great advantages in building a new house is that you can lay out the HVAC and other mechanical systems for easy accessibility. Provided it's done correctly, I like the idea of creating a mechanical room in which all of the HVAC and water-heating equipment is located, along with the central vacuum, freezer, and circuit breaker box. This isn't just a matter of convenience and accessibility, it also makes it possible to put all of the ugly, noisy appliances together in one room — a room that's tucked out of the way and appropriately soundproofed. If combustion appliances are going to be placed in the mechanical room, ventilation needs to be considered right from the very start, and then reevaluated whenever there's an equipment change.

Another smart idea if you're building new is to position the kitchen and baths so that pipe runs are as short as possible. Lengthy pipe runs increase plumbing costs and result in wasteful and inconvenient time lapses as hot water has to travel between the water heater and distant taps.

Integrated Systems

Simply put, integrated systems combine space-heating and water-heating functions into a single system. Instead of having two fuel lines, two burners, and two flues, the equipment is consolidated with the idea of reducing costs and improving efficiency.

Early-day boilers designed for space heating were among the first candidates for integration. By the simple and inexpensive means of running a copper coil through the middle of the boiler pot, and drawing potable water through it, designers found a way to produce hot water for the home as a co-product of space heating.

Though no one manufactures them anymore, boilers with "tankless coils" are still fairly common today, particularly in New

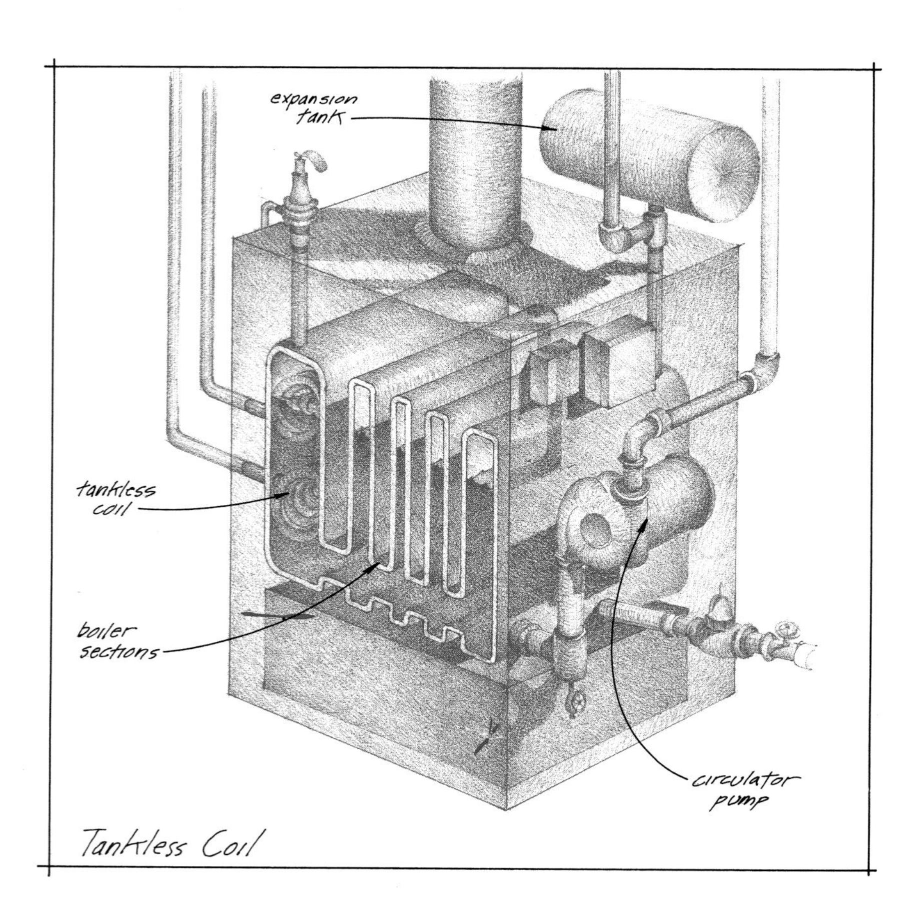

Tankless Coil

England. The problem with tankless coils is that you have to keep firing the boiler all summer long just to get hot water for your taps and showers. This is akin to boiling a whole lobster pot full of water just to make yourself a cup of tea.

What we have done on some "This Old House" projects is to disconnect the tankless coil and install a superinsulated hot water storage tank alongside the boiler. When the water temperature inside the storage tank falls below its setpoint, the boiler comes on and circulates hot water through a heat exchanger coil inside the tank. The boiler still has to work a little through the summer — even on the Fourth of July — but the storage tank enables it to fire a lot less often and to use appreciably less fuel.

An indirect storage tank can be added to virtually any kind of boiler, be it old, young, or middle-aged. But bear in mind that the higher the efficiency of the boiler, the less you're going to pay for hot water.

The best candidates for indirect water heating are people who have a relatively large demand for hot water, who already have a boiler installed for space heating, and who plan to live in their house long enough to recoup the investment.

While an indirect tank costs more up front, it will almost always save you money versus a tankless coil, and is usually a bargain compared to conventional storage-type water heaters too. (See the cost comparisons listed on page 185.)

Over the years I've used different makes of indirect tanks, including Amtrol (plastic tank with copper-fin coil), Phase III (stainless-steel tank-in-tank design), Viessmann (stainless-steel tank with stainless-steel nonfinned coil), and Super-Stor (stainless-steel tank with cupronickel finned coil) — all with good success.

With any make, it's important that the tank be designed for easy cleaning, since the coils — especially fin-type coils — can scale up over time.

Another cost-effective way to integrate space and water heating was discussed and illustrated in chapter 6 (see pages 103–104). Using that approach, hot water from a high-efficiency water heater or boiler is piped through a coil inside an air handler. As indoor air is fan-blown across the hot coils, it's heated, and then distributed into the house through conventional ductwork. In that kind of setup, the efficiency of the space heating depends in large part on how efficient the water heater or boiler is. Mor-Flo Industries, AMTI Heating Products, GlowCore-MTD, and Apollo Comfort Products, a division of State Industries, are the leading manufacturers of this type of system.

In my view, integrated systems are the irresistible wave of

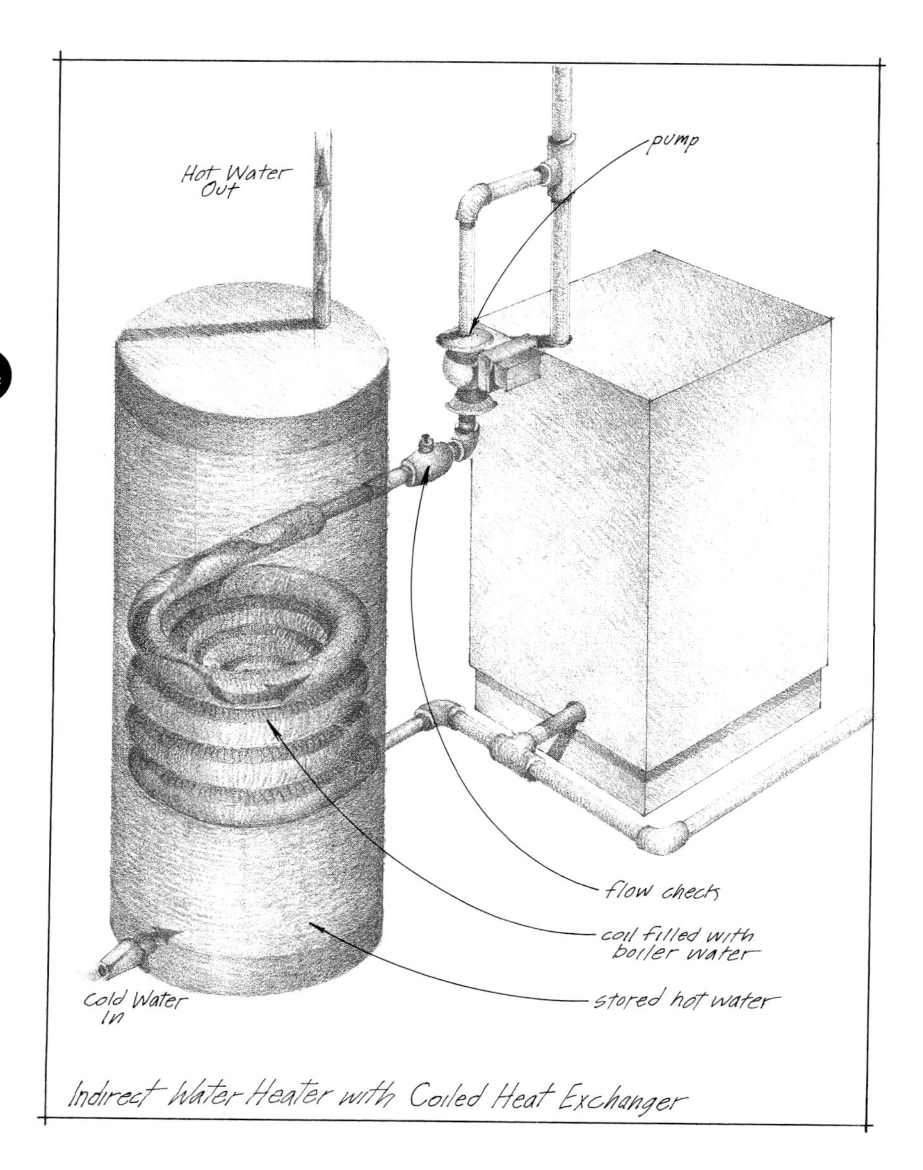

Indirect Water Heater with Coiled Heat Exchanger

the future in HVAC engineering. In addition to saving energy, integrated systems also conserve space, which is an important consideration in modern homes.

I expect in my lifetime to see an integrated machine that can handle a home's space heating, ventilation, air conditioning, and hot water needs in a single integrated package. The idea isn't as far out as you might think — some of the three-way heat pumps already on the market are only a short step or two away from achieving it. We'll talk more about integrated heat pump systems in chapter 11.

Heat Pump Water Heaters

Like other types of heat pumps, heat pump water heaters use a refrigeration cycle — much like the one that keeps your refrigerator cold — to move heat from one place to another. In this case, heat is taken from the surrounding air and pumped into a hot water storage tank.

Though a heat pump water heater typically costs more than twice as much as a high-efficiency electric water heater, it uses only one-third to one-half as much electricity to operate. For a family of four, the *life-cycle* cost of a heat pump water heater would be about $3,850 — not as cheap as a gas water heater to be sure, but a savings of more than $2,200 over a high-efficiency electric model.

Heat pump water heaters are best suited for hot, humid climates because they help cool and dehumidify the air inside the house as a cofunction of heating water. Inversely, they lose some of their appeal in colder climates because they pull heat out of the house, which has to be replenished. Sometimes it can make sense, as in a recent installation I saw in Germany, to locate the heat pump water heater in the attic or other unconditioned space.

Some heat pump water heaters are sold with their own integrated tank, a nice option for new construction or for replacing a conventional water heater. Other models are designed to be retrofitted onto an existing tank.

The Consumer Guide to Home Energy Savings, published by the American Council for an Energy-Efficient Economy, lists Reliance, State, and Therma-Stor as the most efficient heat pump water heaters on the market. By the way, the guide includes valuable performance information on many other types of water heaters too, as well as boilers, furnaces, and air conditioners. (Write: American Council for an Energy-Efficient Economy, 1001 Connecticut Ave., N.W., Suite 535, Washington, DC 20036.)

Solar Water Heating

Like most people, I love the *idea* of heating my water with solar energy. Not only is it free and renewable year after year — as long as the sun keeps shining — but it avoids all the nasty environmental and economic consequences that come with producing, importing, transporting, and burning fossil fuels and reacting uranium.

Of course, loving the idea and executing it in a practical manner can be very different.

The solar industry has gone through several periods of renaissance and decline since the turn of the century. The great boom in solar R&D and sales that started in the late 1970s went bust in the mid-1980s after the Reagan administration killed the federal tax credits for solar. Today, the solar water heating industry is a lot smaller, to be sure, but it's also more stable, competent, and honest. And it's growing again.

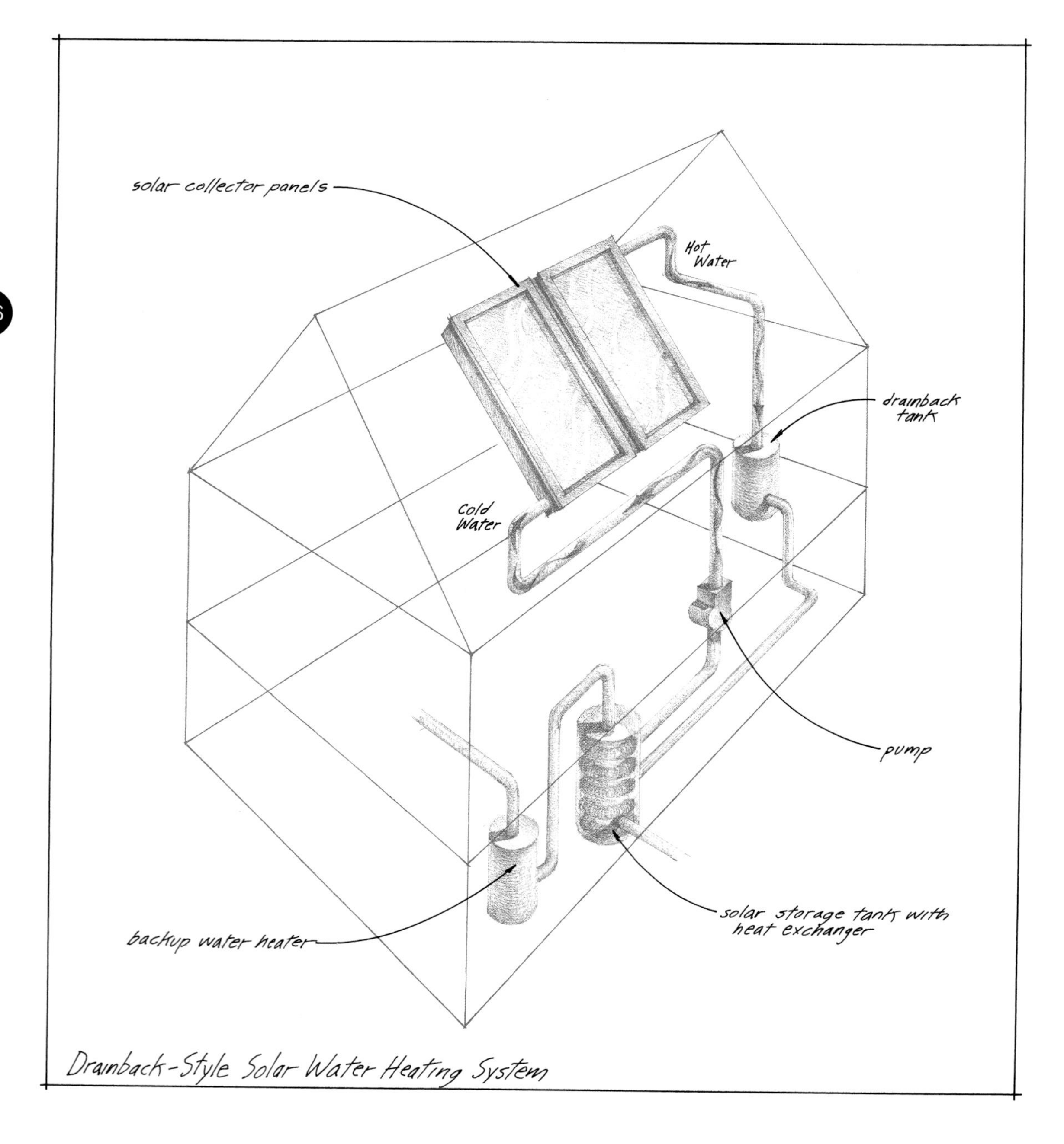

Drainback-Style Solar Water Heating System

Many of the surviving companies are manned with true veterans, who have weathered hard times and know their stuff. The R&D carried out in the 1970s and '80s has helped make today's equipment more affordable, efficient, and reliable. Pumped, flat-plate collector systems that might have cost $6,000 or $7,000 in the early 1980s are being installed today for under $3,500.

As I mentioned in chapter 5, the solar water heating system we built for Justin and Genevieve Wyner proves that solar can be done right. Located in Brookline, Massachusetts, the system has

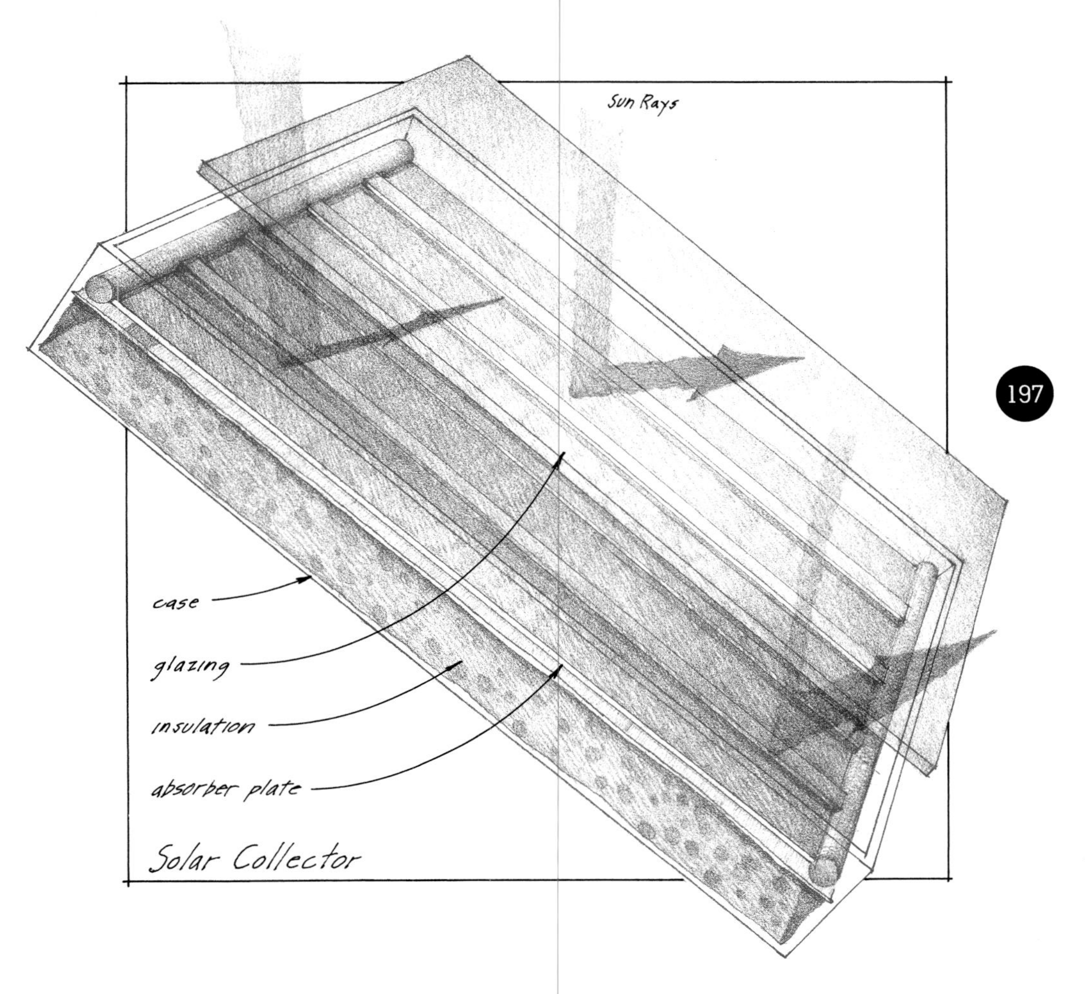

been working dependably since 1984, delivering about 80 percent the Wyners' year-round hot water needs.

That design — illustrated here — is fairly typical of solar water heating systems that have been installed in New England, with one exception. The Brookline house has 140 square feet of solar collectors, which is much larger than average. A more typical setup might have 64 square feet of collectors (two 4x8 panels), supplying 50 to 60 percent of the home's annual hot water needs.

The site-built collectors we used in Brookline are essentially tight, insulated boxes with a top plate of clear glass. Inside each box is a black copper absorber plate that has small tubes running through it to conduct the working fluid — in this case water. Sunlight passing through the glass strikes the absorber plate and is converted to heat. Since the heat can't readily escape back through the glass, it's trapped inside, much the way that heat gets trapped inside your car when it's left in a sunny parking lot with the windows up.

In the Wyners' system, water is pumped through the collectors, where it picks up heat, and then routed into a solar storage tank. A second loop of pipes, fitted with a pump and heat exchanger, moves heat from the solar tank into the hot water tank. During cloudy periods, when there's not enough solar energy to meet the Wyners' needs, a backup electric element in the hot water tank kicks in.

Freeze protection — a key element in solar design — is accomplished by means of a sensor located inside the collector. When the temperature there drops below 40°F, signaling the danger of a freeze, the controller drains all the water out of the collectors into a holding tank. That's how the name "drainback system" was coined.

Solar collectors can also be protected against freezing by filling the solar side of the loop with antifreeze. This type of design, know as a "closed-loop antifreeze system," is also common in New England and other northern climes. Its biggest drawback, vis-à-vis the drainback system, is that antifreeze can't hold as much heat as water and has to be replaced every few years.

Because I'm a firm believer in the KISS principle (Keep It Simple, Stupid!), I've always preferred solar water heaters that are self-pumping. These simple designs — known as "thermosyphon" or "passive" water heaters — don't require electric pumps or sophisticated controls, so there's a lot less to go wrong. When sun shines on the collectors, the system simply pumps itself, because warm water rises and cold water falls.

The catch, of course, is that most thermosyphon systems can't work unless the storage tank is located *above* the collectors. That means locating the collectors somewhere down-slope from the house or, if you buy a system in which the collector and tank are integrated, having the storage tank up on the roof. The first option isn't practical for most home sites; the second puts a big, usually ugly tank full of water up on the roof where it's subject to freezing and high winds.

But a new type of passive water heater — called the Copper Cricket — has overcome these drawbacks. The Cricket, manufactured by a Eugene, Oregon, company called Sage Advance, has no pumps, valves, or controls — in fact, it hasn't got any moving parts at all. Yet it's patented percolating action makes it possible to position the collector *above* the storage tank. And since Cricket uses a water-alcohol mixture as its heat transfer fluid, the system is guaranteed against freezing even in the toughest February. Moreover, the system can be easily retrofitted to work with a conventional electric water heater.

Tests run by the Oregon Department of Energy and the Solar Rating and Certification Corporation rank the Cricket as one of the most efficient solar water heating systems on the market. Here in Boston, a Cricket with a single 4x10-foot collector would provide about 45 percent of a family of four's year-round hot water needs. In sun-rich Miami or Phoenix, the system could meet 80 to 90 percent of those needs. A do-it-yourself Cricket kit sells for about $2,200, will pay for itself in saved energy in two to ten years, and should last the life of the house, the company says.

If you think solar water heating might be a practical investment for you, here are some pointers to help you decide.

- While solar water heating can work even in the extreme north, it's cost-effectiveness improves the farther south you go.
- The larger your family and hot water load, the more cost-effective solar becomes.
- Solar is generally more cost-effective when it's displacing electric water heating than gas.
- Solar water heating is most cost-effective when the collectors are sized to provide one-half to three-quarters of the hot water load. Beyond that, adding more collectors may mean diminishing returns on your investment.
- Solar water heating is especially cost-effective when the system is built into a new home and financed as part of the original mortgage.
- The roof surface on which the solar collectors are to be mounted should be oriented within 35 degrees of south. The collectors need to be largely unobstructed by evergreens or other sun-blocking objects. Avoid bat-wing and other ugly collector configurations that can detract from the looks and value of your house.
- Some state and city governments offer tax breaks and other incentives for solar. Find out if any of these might apply, and see if your local utility has any pro-solar programs.
- As with other HVAC installations, an experienced and conscientious installer is the most critical ingredient.

Two good sources for additional information on solar water heating are the Solar Energy Industries Association (777 North Capitol Street, N.E., Suite 805, Washington, DC 20002; [703] 524-6100) and the Florida Solar Energy Center (300 State Road 401, Cape Canaveral, FL 32920; [407] 783-0300).

Instantaneous Water Heaters

If you've ever traveled abroad, you've probably seen lots of instantaneous water heaters. They're usually mounted on bathroom and kitchen walls close to the point where hot water is used. In fact, some people call them "point-of-use water heaters."

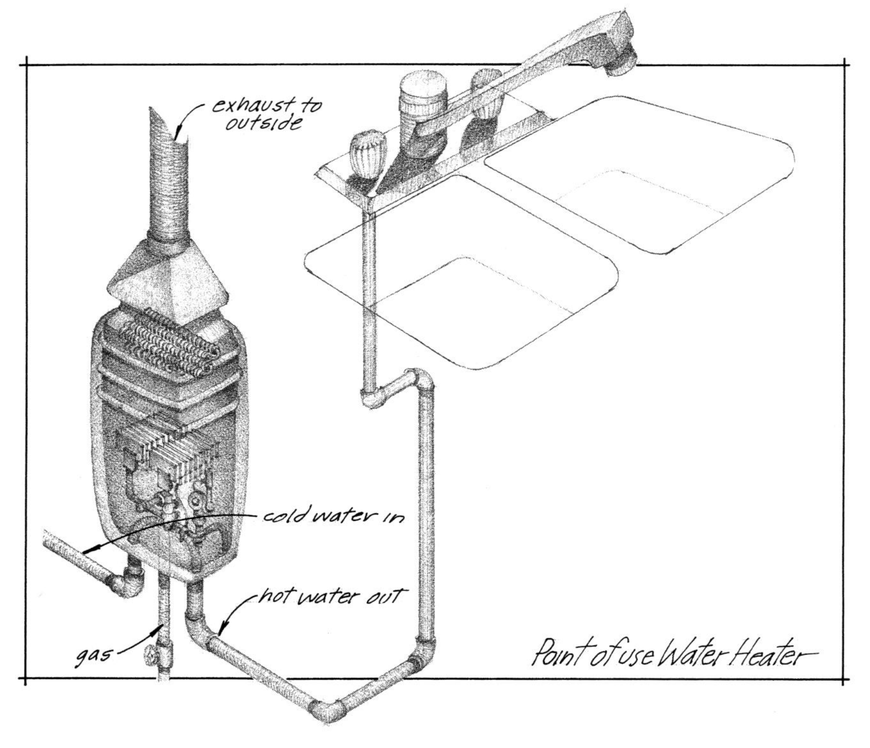

Instantaneous water heaters are popular in Europe and elsewhere around the world because they take up so little space, have a long life, don't have any standby heat losses from a big tank (because there is no tank), and can produce hot water indefinitely as long as the flow rate is kept within certain limits.

Gas-fired models offer the additional advantage of having very low operating costs and among the lowest life-cycle costs of any water-heating option available.

Electric models, on the other hand, can draw frightening amounts of power — as much as 45 amps at 220 volts. If these were ever to catch on in New England in a big way, we'd have to start planning for Seabrook reactors 2, 3, 4, and 5!

Though instantaneous water heaters have been sold in the U.S. for years, their market niche remains small. The chief drawback, with both gas and electric models, is that their flow rate is sharply limited.

When you open a tap, the flow of water activates the gas burner or electric coil inside the heater. Water coursing through the heat exchanger is heated as long as the tap remains open. When the water is turned off, it deactivates the burner or element. The faster the flow of water, the lower its temperature.

Flow rates on gas models range up to about 3 gallons per minute, while the best electric units deliver only 1.2. No single gas or electric model on the market has enough flow to satisfy the large and simultaneous demands that American families typically place on their hot water systems. The only way to really meet those kind of demands with instantaneous water heaters would be to install multiple units around the house, which, in the case of gas, would mean multiple fuel lines, burners, and flues.

Europeans and other people around the world have been generally willing to conserve water and to manage their water-heating loads to an extent that makes point-of-use water heaters practical for them. Most Americans, having grown up with big hot water storage tanks in the basement, aren't prepared to curtail or schedule their relatively lavish use of hot water.

Though you may not have any use for an instantaneous water heater as a primary system, they can make good sense in a cabin or vacation home, for a room addition or in-law apartment that's far removed from your home's main water heater, or as a complementary system to solar water heating.

If you go shopping for one, be forewarned that a lack of standardized ratings in the industry make product comparisons difficult.

Gas-fired instantaneous water heaters are sold under various brand names, including Aquastar, Bosch, Bradford-White, Fastomatic, Myson, Paloma, and Vaillant. Among the electric models are Acutemp, Chronomite, Instantaneous, and Powerstream.

●

Fresh Air, Clean Air, Cool Air

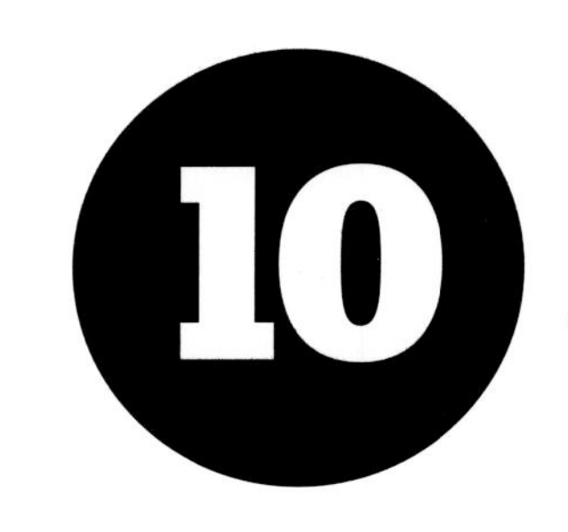

On days in July and August when it feels hot enough to blister the paint off the side of your car, and humidity rises to the point where you feel more like a hot mop than a human being, my strategy for comfortable and cost-effective cooling starts with the same three-step principle that I use for heating. First, do all you can through conservation, passive techniques, and lifestyle changes to cut your load. Second, maintain the equipment you've already got. Third, if the first and second steps don't give you enough comfort or economy, shop carefully for new equipment.

Defending Your Home against Heat and Humidity

The wonderful thing about having a tight, well-insulated home is that it works hard for you all year round. Having your doors and windows properly caulked and weatherstripped serves not only to keep out cold, dry air in the winter but also to seal out hot, humid air in the summer. Likewise, the insulation you lay down in the attic works in your favor summer and winter alike. (Even in a hot climate, where there's little call for space heating, experts recommend insulating the attic to R-30 to conserve air conditioning.)

Tightening and insulating your house probably won't eliminate the need for mechanical cooling, but it *will* enable you to use less, which will reduce your monthly utility bill and extend the life of the equipment. And when it does come time to junk the old air conditioner, you'll be able to buy a smaller, less expensive replacement.

Another important way to cut your cooling load is to protect your windows against incoming heat. Studies show that as much as 25 percent of the heat inside a house comes streaming right in through the glass.

Trees, shrubs, and vines are among our most powerful allies when it comes to summertime shading. Carefully placed vegetation can shield your roof and walls — and especially windows — from both direct and indirect solar gain. In fact, a single mature tree has the cooling power of five room air conditioners (about 60,000 BTUs). Trees and shrubs also lower the air temperature around the house and enrich the air with oxygen.

Hopefully, the people who laid out the site plan for your house were smart enough to preserve all the trees and shrubs they could, especially on the east and west, where the summer sun can really bear down. Where the existing vegetation was lacking, perhaps they took the time and care to do some thoughtful landscaping.

Unfortunately, that kind of building scenario is pretty rare. More typical are the home sites that were clear cut and bulldozed without a second thought, because it was quicker and cheaper that way. If the land around your house falls into the "denuded" category, don't lose heart. It's never too late to do some intelligent landscaping. Some varieties of trees and shrubs — and especially vines — will surprise you at how fast they grow. The least expensive way to add a tree or two is to buy saplings from your local nursery and plant them yourself. Or, if you have the budget, you can have more mature trees trucked in and planted by pros.

Here in New England, it's usually good strategy to place shade trees to the east and west sides of the house, but to leave the southern exposure open so that it can receive wintertime sun. A line of evergreen trees or thick shrubs along the northern side of the house can help shield it against the winter wind.

Obviously, the type of vegetation you select and where you place it will depend heavily on your climate, how your house is oriented on the site, and the position of windows, doors, and other design features. Your local nursery may have people who understand how to use vegetation for passive cooling. Or if you have the extra budget, you could hire a professional landscaper or landscape architect.

In addition to the cost and comfort advantages you'll enjoy from natural shading, trees and shrubs will make your home more attractive and increase its resale value.

Where vegetation can't do the job, you'll have to rely on some type of man-made shading. Adding a new porch, carport, or trellis to the sunny side of a house is one way to shade and buffer it from the summertime sun. Another is to extend the existing eaves so that they'll provide more ample shade to the windows below.

Window awnings are among the most effective sun-blockers you can buy. They come in a wide variety of fixed, adjustable, and removable configurations, and an even wider variety of materials and colors.

If awnings aren't to your liking, you could choose exterior roll screens, louvered shutters, or dense mesh insect screens. All of these share a common trait: They block the sun before it hits the glass.

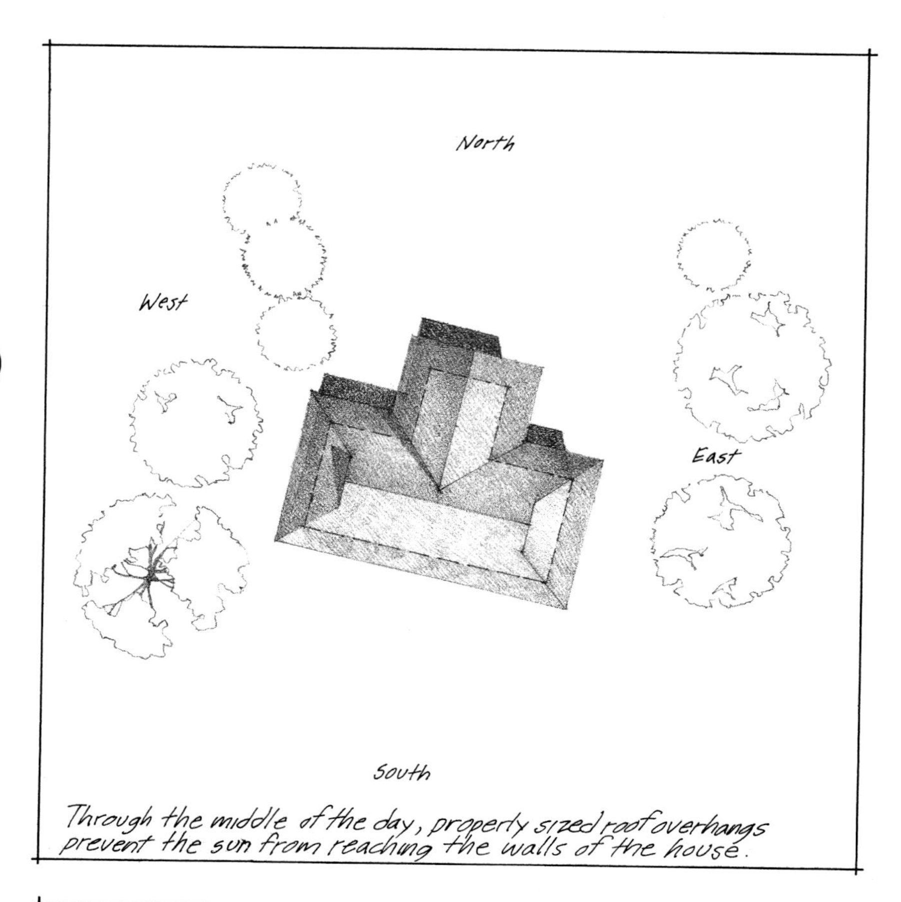

Through the middle of the day, properly sized roof overhangs prevent the sun from reaching the walls of the house.

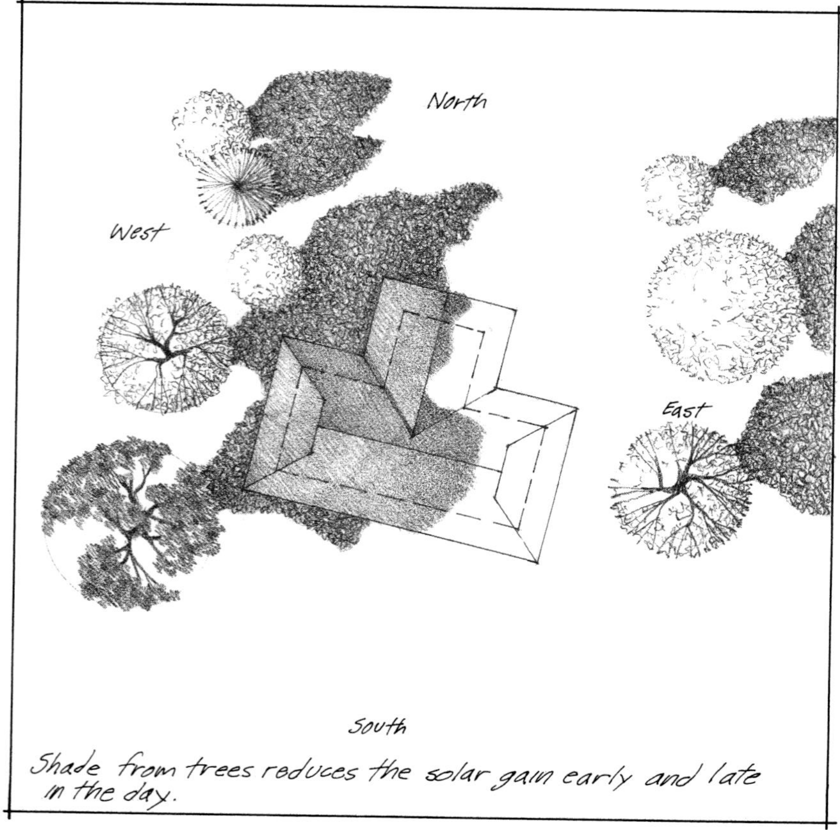

Shade from trees reduces the solar gain early and late in the day.

How to Control Moisture in Your Home

If your house has too much moisture in the air, molds, mildews, and bacteria can find a ready breeding ground in carpets and on wall and ceiling surfaces where water vapor condenses. Moist air can also seep *through* the wallboard, condensing on cold studs and joists, and that sets the stage for rot.

Too *little* moisture in the air can also create problems. When the indoor humidity is too low, people's throats and nasal passages may become irritated, furniture joints can come unglued, book bindings crack, paint may not hold to some surfaces, wood floors shrink, and static electricity can become a nuisance.

If you want to get a better measure of indoor humidity, or monitor changing conditions, you can borrow or buy a hygrometer. Radio Shack sells a wall-mounted unit (for about $30) that measures and displays both the relative humidity and indoor temperature. More precise and expensive models — called sling psychrometers — are also available.

If you find that you do have a problem, here are some effective steps you can take to raise or lower indoor humidity.

To Lower Indoor Humidity

- Use more natural ventilation. This is done simply by opening windows and doors to let your house breathe.
- Get rid of some of your house plants.
- Remove any firewood that's stored inside the house.
- Don't let showers run any longer than necessary or cooking pots boil needlessly.
- Add mechanical ventilation. This might include (installed singly or in combination) a bathroom exhaust fan, power vent over the range, window fan, whole-house fan, exhaust-only ventilation system, or an air-to-air heat exchanger. (See the text for details.)
- Install air conditioning. Whether it's a window unit or central system, a properly sized air conditioner dehumidifies the air as a function of cooling. Some air conditioners have a ventilation-only cycle that can help control indoor humidity and save you money on cooling costs.
- Add a mechanical dehumidifier. This may be a poor compromise, since the dehumidifier raises the air temperature as it removes moisture.

To Raise Indoor Humidity

- Adopt some new house plants. They'll add both oxygen and water vapor to the air.
- Set a kettle or pan full of water on top of a radiator or wood stove.
- Cut back a little on the use of bath and kitchen exhaust fans, or other mechanical ventilation.
- Tighten up the envelope of your home with caulk, weatherstripping, insulation, and/or better-quality windows. Since your house will breathe less following these alterations, more moisture will remain trapped inside.
- Buy a humidifier. You can purchase a room model (designed for stand-alone use) or have a permanent humidifier installed in the ductwork on your forced air furnace. The better models usually have a built-in humidistat, which senses indoor humidity levels and controls the unit accordingly. Be forewarned, however, that humidifiers can use a lot of electricity and must be rigorously cleaned and maintained so that they don't become a breeding ground for molds, mildews, and bacteria.

There are a couple other ways to raise humidity inside a house, but both of them, in my view, come with serious drawbacks. The first, for homeowners who use a fireplace or wood stove, is to store some of their green firewood indoors. As the wood dries, it will release moisture into the air. Unfortunately, the firewood may also release a swarm of spiders and carpenter ants, so I *don't* recommend this approach.

Another method — much advertised in recent years — is to redirect the warm air vent on the clothes dryer so that it exhausts indoors. While this reclaims a lot of waste heat and humidity from the clothes dryer, it usually pumps *too much* humidity into the house and can fill the air with dust and lint. Special vent diverters are advertised and sold for this purpose, but I'd steer clear of them.

Reflective window films are in a class by themselves, since they're mounted directly on the inside of the glass. Their reflectivity — that is, the amount of incoming heat they can block — can be as high as 90 percent. But applying these thin adhesive films to the glass can be tricky (it's best left to a pro) and some films leave the glass with a noticeable tint. Moreover, I'd be leery of using film on

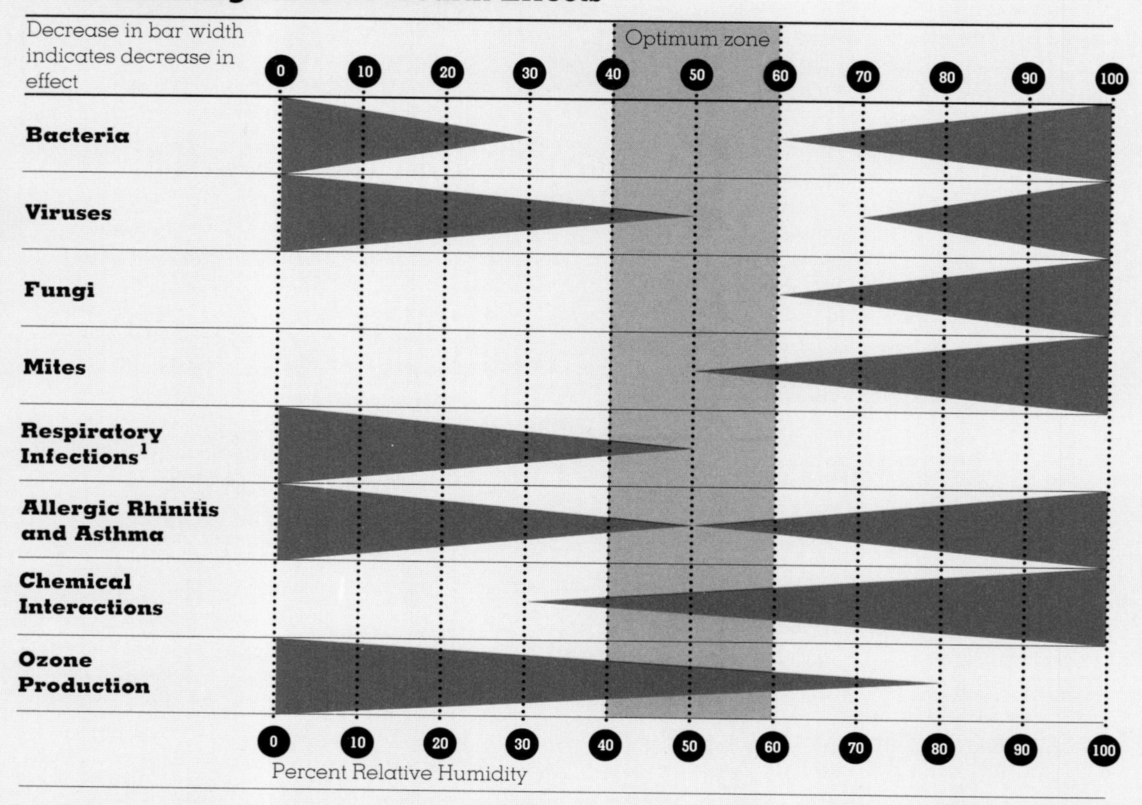

Insufficient data above 50 percent relative humidity
SOURCE: Theodor D. Sterling and Associates, Ltd., Vancouver, B.C.

double-pane glass — especially on east- and west-facing windows — since it can heat up the glass to the point where it breaks.

Finally, you can choose among a wide assortment of shades, curtains, and blinds that are designed to hang *inside* the window. These interior treatments can help — especially if they have a white backing — but aren't nearly as effective at rejecting heat as treatments that shade the window on the outside.

As a rule, *versatility* is the magic word when it comes to controlling the sun. In a world of changing seasons and unpredictable weather, homeowners are best served by window treatments that are adaptable.

No matter what type of shade you buy, make sure that it's durable (especially if it's going outside) and easy to use (otherwise you probably won't use it).

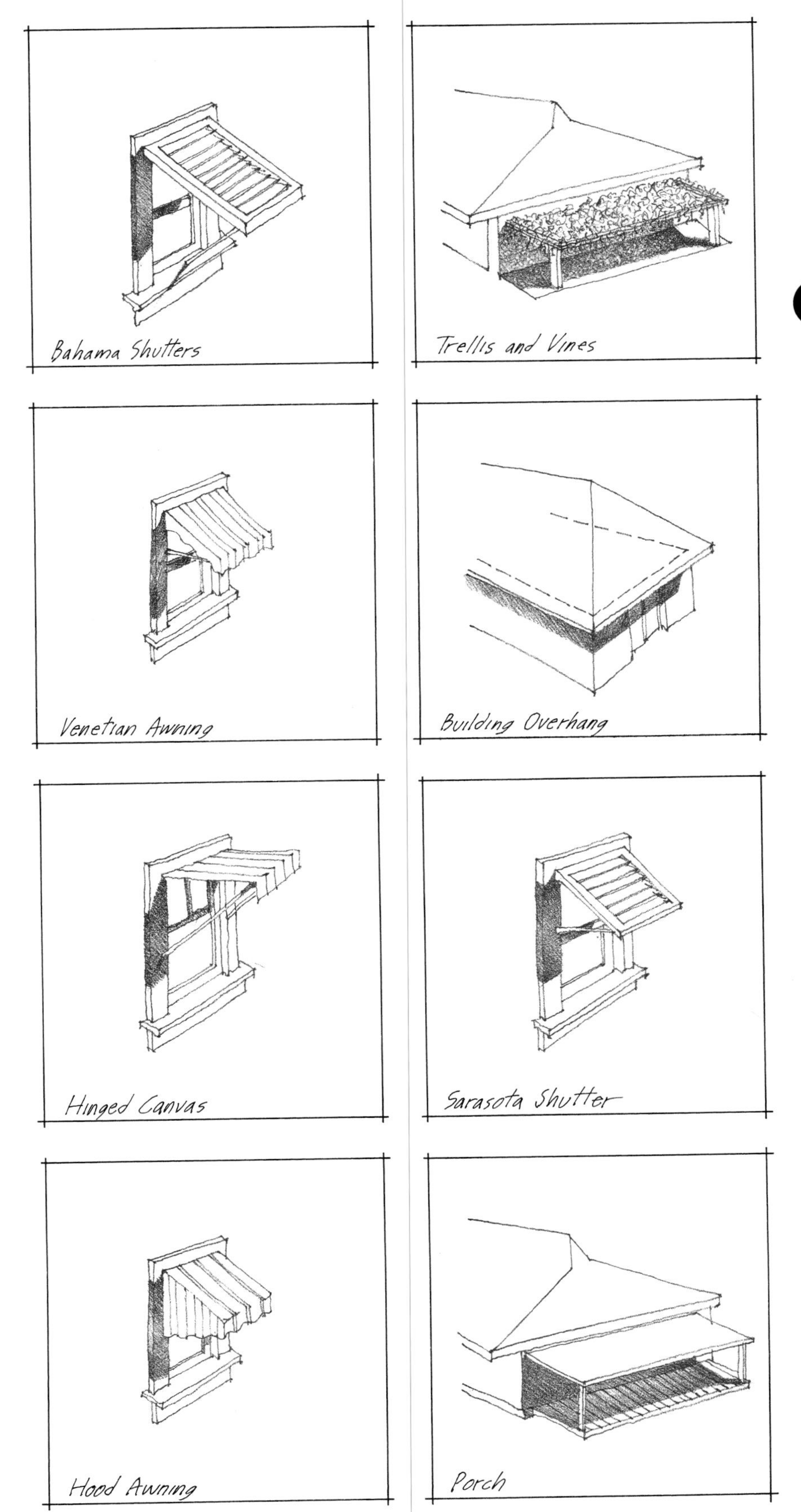
Bahama Shutters
Trellis and Vines
Venetian Awning
Building Overhang
Hinged Canvas
Sarasota Shutter
Hood Awning
Porch

Installing a Radiant Barrier

Even if your attic is well insulated and has plenty of ventilation, a lot of summertime heat can still radiate down into the rooms below.

A simple and cost-effective way to block much of that heat is to install a radiant barrier on the underside of the roof decking. All you need for the project are some rolls of reinforced reflective foil, a staple gun, and the stamina to spend a few hours up in your attic.

As shown in the drawing, the foil is stapled to the rafters with its shiny side facing down — that is, toward the attic floor. The resulting barrier, covering the entire underside of the decking, will block heat that would otherwise radiate down into the insulation, through the attic floor, and on into the house. Bear in mind that the foil doesn't have to be perfectly installed to work well — a few gaps along the seams or even a hole or two in the foil itself won't appreciably cut its effectiveness.

If you live in the Sunbelt, this simple do-it-yourself project can trim 8 to 12 percent off your annual air conditioning bill. It may save even more if you have air conditioning ductwork running through your attic.

The farther north you live, the less cost-effective radiant barriers become. I wouldn't recommend them at all in New England, because our air conditioning season here is so short and the barrier has little value during the heating season.

Your out-of-pocket costs for the project will run 10 to 40 cents a square foot for the reinforced reflective foil. (Double-sided foil — that is, foil that's shiny on both sides — isn't worth the extra cost.) Hiring a contractor to do the installation will probably add 15 cents a square foot to the project's cost.

If you decide to do the work yourself, and have fiberglass insulation in the attic, don't forget to wear a dust mask and protective clothing.

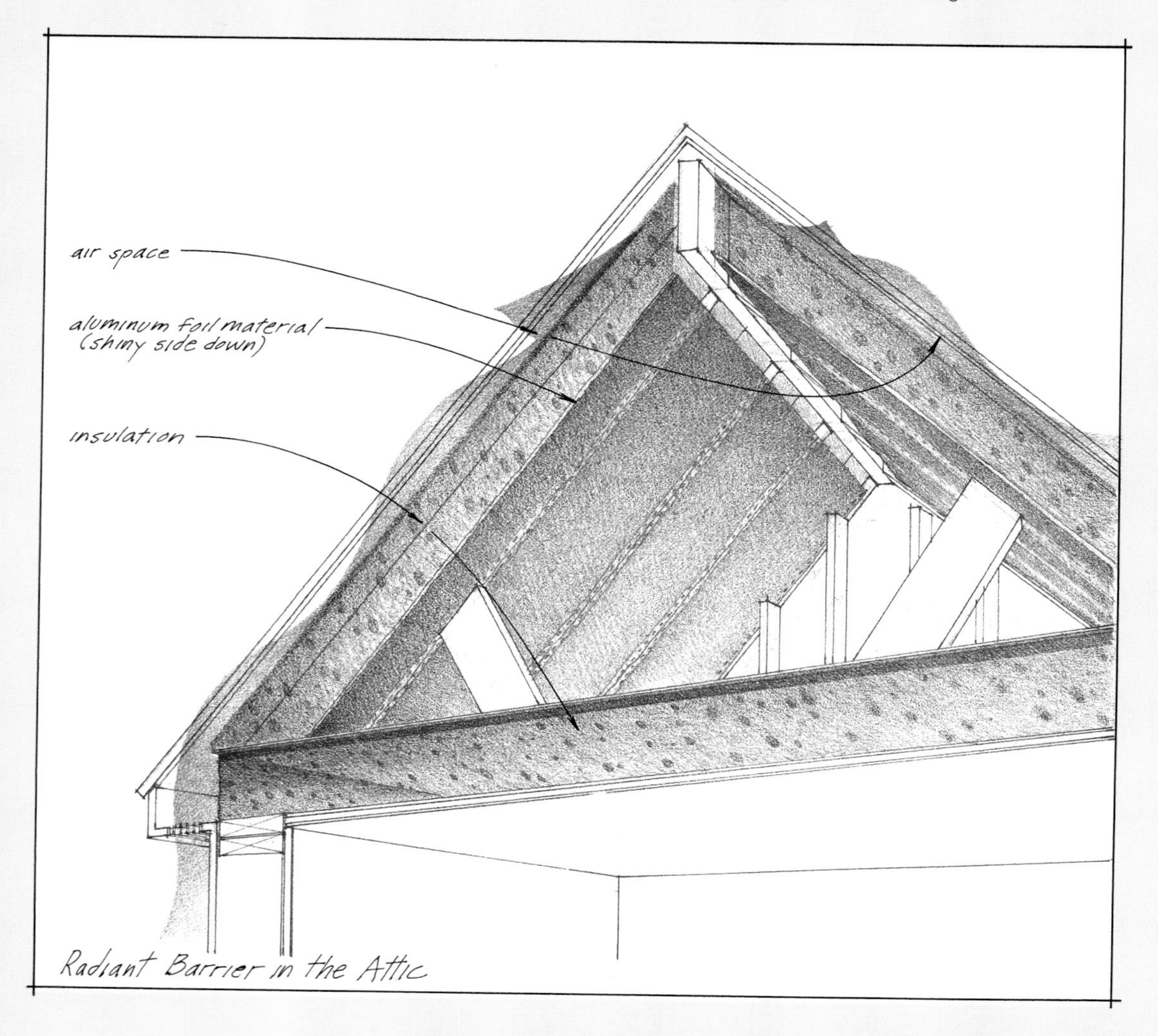

Radiant Barrier in the Attic

Natural Ventilation

It's ironic that one of the most important elements in creating a comfortable, healthy, and affordable home — *ventilation* — is often ignored or left to chance. Without adequate ventilation, heat, humidity, air pollutants, and odors build up inside our homes, taxing both our budgets and our bodies.

A good first place to focus your attention is on the attic, where summertime temperatures can reach 150°F! It doesn't take a Ph.D. to figure out that if the attic isn't properly ventilated, that pent-up heat is going to radiate and conduct its way right down into your home — with profoundly unhappy effects on your comfort and air conditioning bill. (The penalties will be even more severe if there's air conditioning ductwork running through the attic, and worse still if that ductwork is uninsulated and/or leaking.)

By ventilating the attic, you also help to keep it dry, protecting the joists, rafters, and sheathing from rot and the shingles from premature failure.

In the wintertime, attic ventilation (along with insulation) will also help discourage the formation of ice dams on the roof, since snow is less apt to undergo rapid freeze-thaw-freeze cycles when the attic and outdoor temperatures are close to equal.

The goal, all year long, is to keep the temperature inside the attic as close as possible to the outdoor temperature. To accomplish this, the attic should have one to three square feet of vents for every 300 square feet of roof area. Since most attic vents are protected with screen wire and louvers, which impede the flow of air, the total vent area has to be reckoned according to the materials you choose. Here are some guidelines (courtesy of Boston Edison's Energy Crafted Home program) that will help you size the vents correctly.

strip vent

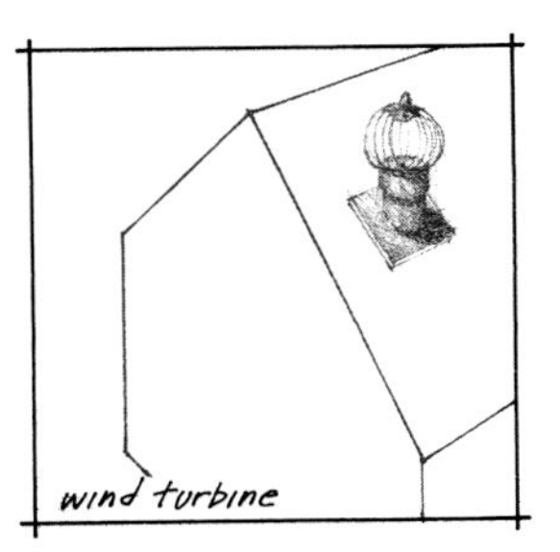

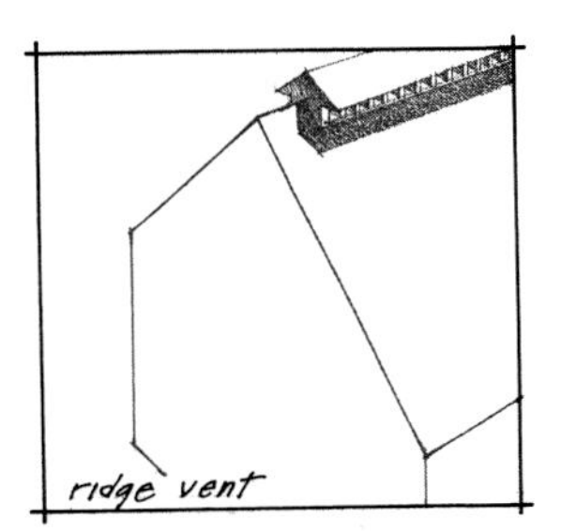

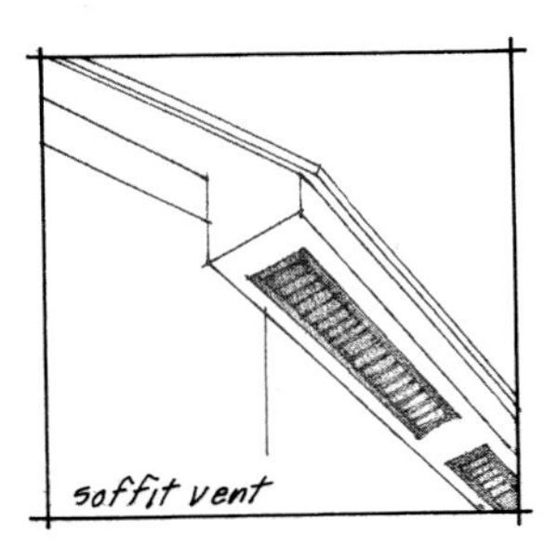

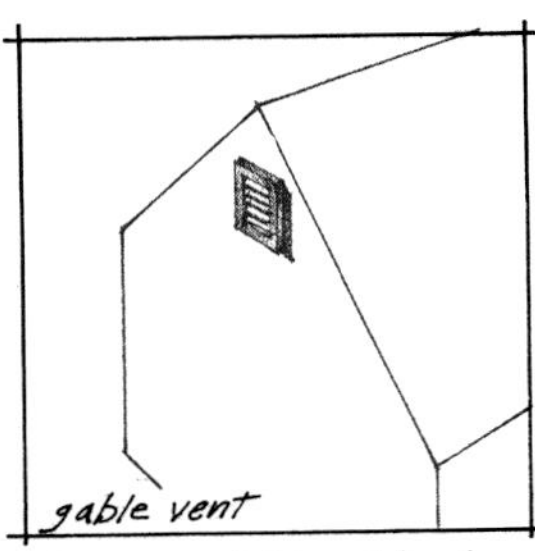

Passive Attic Vents

Attic vents to be fitted with:	Opening required for every 300 sq. ft. of roof area (in square feet)
¼-inch (mesh) screen	1.0
¼-inch screen w/louvers	2.0
⅛-inch screen	1.25
⅛-inch screen w/louvers	2.25
1/16-inch screen	2.0
1/16-inch screen w/louvers	3.0

As long as you meet these standards for natural ventilation, there's no need for power ventilation.

In the drawing, we've shown five different types of passive vents — ridge, gable, soffit, strip (flat), and turbine — though in practice all five probably wouldn't be used on the same house. Turbines are wind-driven ventilators that are most effective in hot, windy climates.

If your attic doesn't have enough vent area, it's a fairly straightforward do-it-yourself job to put in new or larger gable vents or to install soffit vents underneath the eaves. Another nice do-it-yourself job — for those of you who live in the Sunbelt — is to install radiant barriers in the attic. I've spelled out some details on the job in the sidebar.

Whether or not to provide summertime ventilation for a crawl space or unheated basement is a stickier question. Many experts and some building codes call for operable screened vents spaced around the perimeter walls so that moisture, which migrates up from the soil or seeps through concrete walls, can be flushed out with fresh air.

The opposition argues that there's little advantage in venting crawl spaces and unheated basements in the summer, since outdoor air may have a higher relative humidity than what's inside. Moreover, when moist incoming air strikes relatively cool surfaces in the crawl space or basement, condensation occurs. And that, of course, is what you were trying to fight in the first place.

Both camps seem to agree that it's good practice to cover the ground in a crawl space with plastic sheeting, overlapped and taped, to keep moisture (and possibly radon gas) from migrating up out of the soil.

As for installing operable vents, I suggest you contact one or two quality builders in your area and ask them about local codes and practices. If you do install operable vents in your crawl space or unheated basement, be sure they're opened in the summer and closed again when heating season rolls around.

Once you're satisfied with the attic and crawl space (or basement), give some thought to how passive ventilation can make the living areas of your home more comfortable and affordable.

Maybe you're lucky enough to live in a home that was properly sited on the land and designed for natural cooling. Good architects and builders will position windows so that they open to the prevailing winds in summer and send a cross flow of fresh air through the entire house. Some go a step further and build wing walls or plant hedgerows that serve to funnel summer breezes toward the windows.

Even if your house wasn't designed with that sort of sensitivity, you can still take advantage of operable windows — especially at night — to flush out the hot humid air of the day. Come morning, when it starts to heat up outside, you can close the windows and bottle up some "cool" for the day.

A good passive ventilation strategy in a hot, humid climate, where there's not much difference between daytime and nighttime temperatures, is to open a window at the lowest point of the house and one at the highest. This creates a "chimney effect," inducing an artificial breeze up through the house.

Before I move on to mechanical ventilation, I'd like to add just a word here about lifestyles. Some of the easiest and most effective ways of reducing your home's cooling load start with you yourself, in the way you dress and schedule your activities.

I'm not suggesting here that you radically alter your family's lifestyle so that you can save $15 a month on air conditioning bills. But isn't it just common sense to dress cool in the summer, so that your body and your air conditioner won't have to work so hard? (See page 18.)

And why not minimize indoor activities that dump a lot of heat and humidity into the air?

Some electric and water appliances — such as the refrigerator and the shower — don't really lend themselves to any kind of modified summer use. But others — like the dishwasher, hot tub, and oven — can be used sparingly or at night without much sacrifice.

One of the easiest ways to reduce your home's cooling load is simply to turn off electric lights when they're not being used. Try touching an incandescent bulb after it's been on for a while and you can feel for yourself just how much heat it's putting into the air. (By the way, fluorescent lights run a lot cooler, last ten times longer, and use 75 percent less electricity than the incandescent type.) To me, these kinds of lifestyle modifications are easy and natural. I don't know about you, but come the dog days of summer I don't feel much like baking a sixteen-pound turkey for dinner or splashing around in a hot tub. How about a slice of watermelon instead, and a cool place in the shade?

Mechanical Ventilation

While passive cooling and ventilating techniques can go a long way toward keeping your indoor air fresh and cool, they can seldom do the job alone.

Floor model and desktop oscillating fans are a fine and economical tool for spot cooling. The problem is, you have to stay in front of them — and at pretty close range — to get much benefit.

Ceiling fans offer a lot more cooling power and range, and can do double duty in the winter by circulating warm air back down into the room where it's needed.

Like other types of fans, a ceiling fan works to cool *you* —

Trethewey's Tips: Ceiling Fans

- Choose a model that's reversible and has at least three speeds, or, better yet, infinite speed control. This will give you year-round flexibility.
- High-quality ceiling fans feature direct-drive motors with permanently sealed bearings.
- A solid cast-iron housing may cost you more, but it will help keep the motor cool and extend its life.
- Avoid manufacturers who use a one-size-fits-all approach to motors. Motor capacity should increase relative to the fan's diameter.
- Size the fan carefully using the guide below.
- The fan blades should be evenly balanced and pitched at a 10-degree angle. Avoid blades made out of cheap materials that can warp or twist.
- Watch and listen to the fan carefully *before* you buy it. The blades should spin freely with no wobble and very little noise.
- Better-quality models will have a good warranty on the motor, and service centers or agents to handle repairs.
- The fan must be securely mounted into the ceiling joists — not to an electric box or with flimsy metal brackets.
- The fan should have a minimum clearance of 10 inches to the ceiling. With a standard 8-foot ceiling, that still leaves more than 7 feet of clearance between the floor and the fan blades. In a room with sloped or high ceilings, the fan should be mounted 7½ to 8 feet above the floor.
- Remember that you can spend a fortune on bells, whistles, and ornamentation, and still not get any more cooling power for your money.

Largest room dimension	Recommended fan diameter (inches)
12 feet or less	36 or 42
12–16 feet	48 or 52
16–18 feet	52 or 56
larger than 18	Two fans

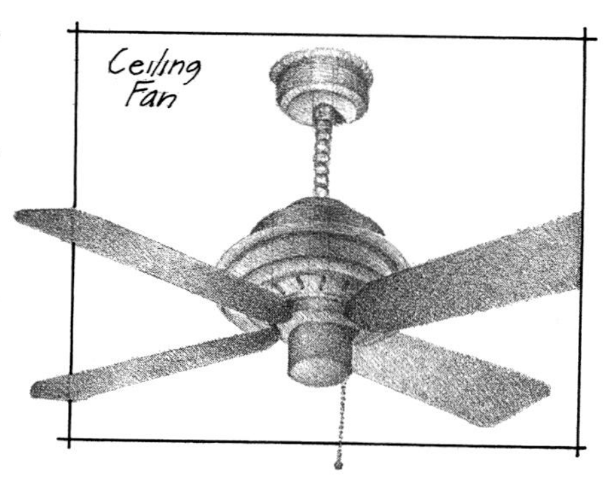

not the surrounding air. It does this by blowing air over your skin, which helps your body shed heat through evaporation and convection. If I were an engineer, I'd be tempted to call this effect an "artificially induced wind chill."

A ceiling fan draws about the same amount of electricity as a 60-watt light bulb, a tiny fraction of the power needed to run an air conditioner. Of course, a ceiling fan can be used in tandem with air conditioning, reducing the overall load and distributing cooled air.

In the winter, ceiling fans are of special value in rooms with high ceilings — particularly cathedral ceilings — where valuable heat can get trapped up high. A very slow (reversed) rotation is preferable in the winter, gently mixing the air without creating drafts. In summer you can turn the fan up high and enjoy the breeze.

To help you in sizing a new fan and shopping for other features, I've put together a list of pointers in the sidebar. As you can see from the pictures we've collected, ceiling fans come in a wide range of attractive housings and colors, with blades made from wood, brass, cane, and other materials.

Bear in mind that small, oscillating fans and ceiling fans are fine for cooling, but aren't much help in bringing fresh air into the house. To do that, in a controlled manner, you need to consider other types of equipment.

Exhaust-Only Ventilation

As the name suggests, exhaust-only systems use one or more fans to blow stale air out of the house.

One very common and fairly inexpensive way of doing this is to install a fan in the bathroom and/or over the kitchen range. These small exhaust fans were introduced back in the 1940s as a means to expel smoke, odors, moisture, and other pollutants from the two rooms that need ventilation most — namely, the bath and kitchen.

One big problem with many of those early-day exhaust fans was the hellacious racket they made. Another was their poorly designed dampers, which leaked even when the fan was off. Fed up with the noise and drafts, many homeowners simply quit using their fans.

Modern-day bath and kitchen fans — at least the better-quality ones — are much quieter and less prone to leak. They're enjoying renewed popularity, especially in new construction, as a least-cost way to meet the new minimum ventilation standards that are being enacted.

Installing an exhaust fan in your bathroom is probably the single most important step you can take toward controlling moisture in your home. (A hood vent over the kitchen range would be a close second.) But bear in mind that these fans aren't designed to cool or ventilate the whole house. Their relatively small motors are meant for intermittent use, not continual, heavy-duty operation. And they only pull a small amount of air. So it's best to think of them as "spot" ventilation, to be used in tandem with other cooling and ventilation gear.

Before you buy an exhaust fan for the bath or kitchen, be forewarned that there are still a lot of cheap, noisy, leaky models on the market that will make you regret the day you bought them. A good bathroom fan is well worth the $75 to $150 it costs. Quality range

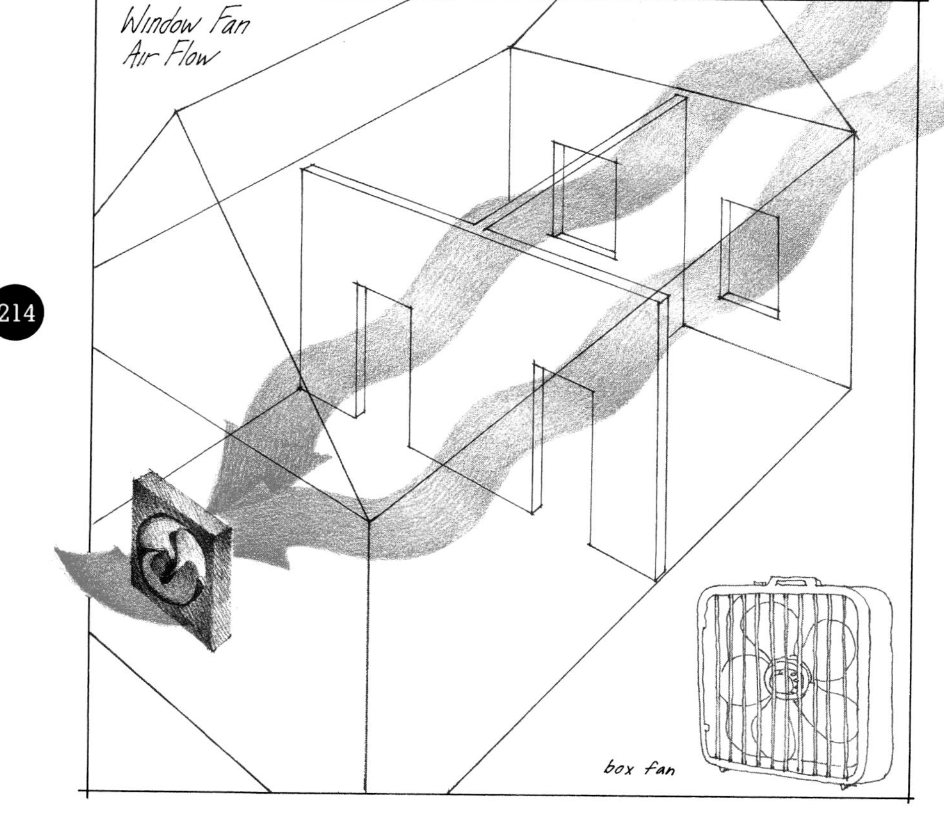

hood fans are $150 to $300. In each case, installation fees will probably add another $100 to $300 to the final cost.

Here are some additional pointers:

- Fans are rated by the cubic feet of air they can move in one minute, or CFM. Exhaust fans for baths typically run from 50 to 110 CFM, while kitchen models are rated from 150 to 600 CFM. Baths or kitchens that have special ventilation needs (a hot tub in proximity, for example) should have a fan with a higher CFM.
- Bathroom fans should have a "sone" rating of 3 or less. (A sone is a standard measure of how much noise the fan makes.) Kitchen fans, which are more powerful, should have ratings of 5 sones or less.
- Make sure the installation has two backdraft dampers. One should come as an integral part of the fan, the other can be installed in the duct or in the weather cap where the duct pierces the wall or roof. (Avoid installing the fan on a wall, since it leaves room for only one damper.)
- Use sheet metal duct that's well sealed and insulated. The duct run should be kept as short and straight as possible.

Corrugated flex duct isn't recommended since it can impede air flow, sag, and collect dirt and moisture.

- Never vent a bath or kitchen exhaust fan into an attic, basement, or crawl space.

Another inexpensive approach to exhaust-only ventilation is to use a simple window fan. Mounted in a shady window away from the prevailing wind, the fan is positioned so that it blows air out of the room. When you turn it on, the fan will create negative indoor pressure, drawing fresh air into the house through open windows or doors on the other side of the house. By leaving an open pathway through the house, you create a stream of fresh, cool air.

Similar to the window fan — but more effective — is the whole-house fan, which is typically mounted in the attic. These powerful fans are up to 36 inches in diameter and designed to pull 3,000 to 7,000 CFM. Good-quality models are built for continuous, heavy-duty use — at variable speeds — and can be surprisingly quiet if the installation is done with care.

Whole-house fans cool and ventilate by drawing hot, stale air up out of the house and exhausting it into the attic. Fresh air, or "make-up air" as it's sometimes called, flows into the lower part of the house through opened windows and doors. By running the fan all night and turning it off in the morning when outdoor temperatures start to rise, the entire house can be flushed out and cooled, buffering it against the daytime heat.

A properly sized whole-house fan should be able to provide 20 air changes per hour (ACH). That means the entire volume of air in your house is flushed out and replaced 20 times in one hour of fan operation.

A good rule of thumb for sizing the fan is to calculate your home's volume and multiply it by .33. For example, a 1,600-square-foot house with 8-foot ceilings would have 12,800 cubic feet of space. Multiplying that by .33 gives you a recommended fan rating of 4,224 CFM. (For fans that carry a "free air" rating — that is, don't take into account the static pressure drop — increase the required CFM by 20 percent.)

While whole-house fans are arguably the best way to cool a house without air conditioning, their effectiveness starts to fall off when the outdoor temperature climbs above 76°F and the relative humidity goes above 60 percent. Nevertheless, a whole-house fan can be a major league money-saver when it's used to complement central or room air conditioners. Since the fan can effectively shorten the season when air conditioning is needed, and reduce air conditioning hours even in midsummer, you can enjoy up to a 50 percent savings on your total cooling bill.

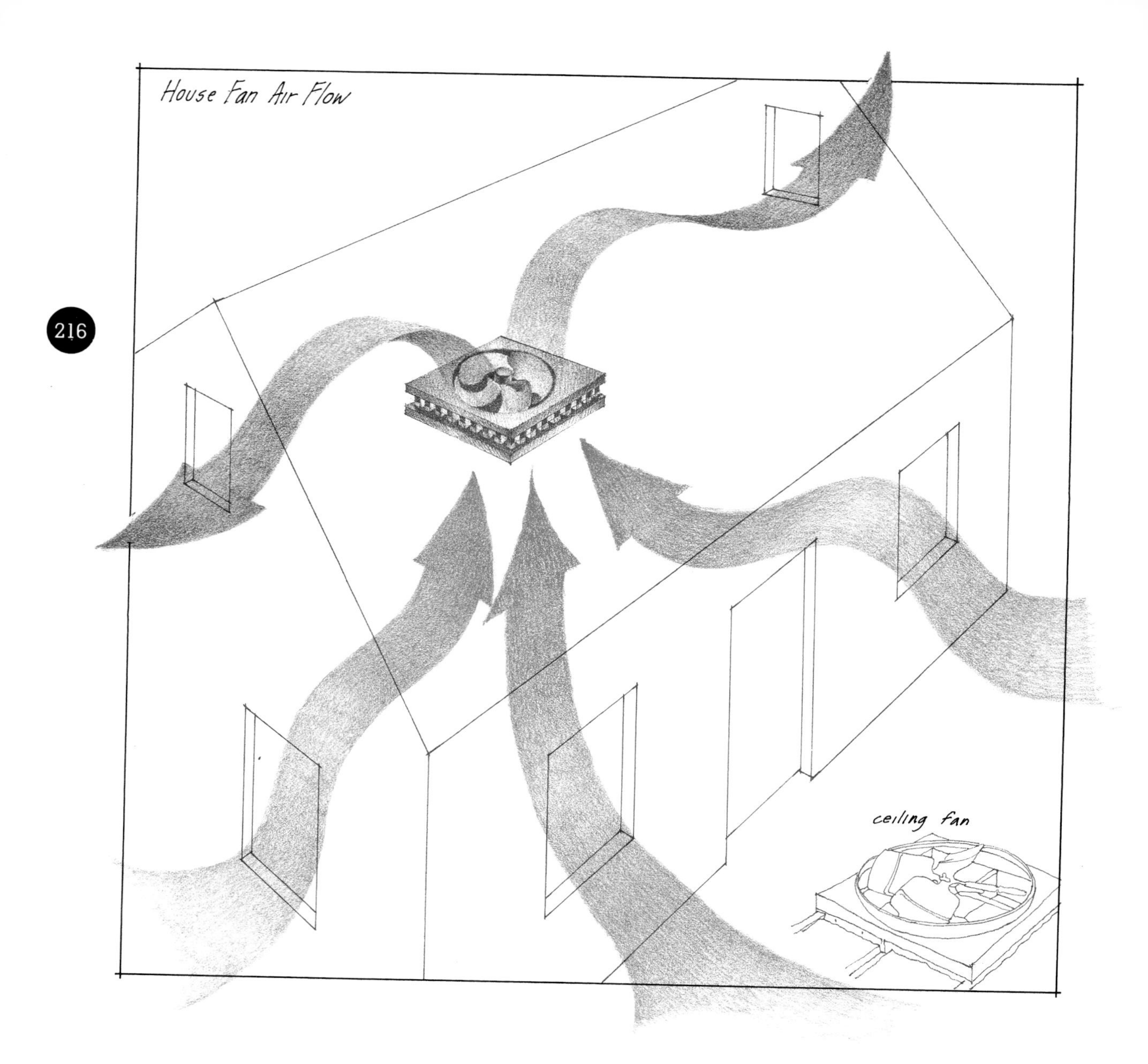

A whole-house fan will cost from $200 to $600 to install, depending on the size and quality of the fan and who does the installation. Whether you do the work yourself or call in a professional, here are some tips for a safe and efficient installation.

- The fan should have variable speeds, with both manual and automatic controls, such as a timer and/or thermostat.
- Be sure to compare the sound ratings (sones) of different fans before you buy.
- Try to meet your ventilation needs with a fan that measures 24 inches or less. That way you won't have to cut ceiling joists during installation.

- Make sure the attic has plenty of open vent area to accommodate the outflowing air. To determine the proper vent area, simply divide the fan's CFM rating by 750 and then factor in the screen and louver adjustments that I outlined earlier. (The fan's exhaust might also be ducted out through a gable end or up through the roof.)
- Remember to open your windows at least three inches to let air flow in. You may want to position your bed near an open window so that you'll be more directly cooled while you sleep.
- Avoid leaving any rough seams or loose fittings in the installation that can cause vibrations and noise. (Some fans come with an acoustic shroud to help dampen noise.)
- Make sure that the fan is carefully weatherstripped and insulated, and that it's equipped with a tight shutter. Otherwise, it will leak warm air up into the attic during the winter.
- The fan should have a fusible link so that it's automatically shut off in case of fire.

Perhaps the greatest drawback to using a whole-house fan is that when cold weather comes, and you have to close the windows and seal off the fan for the winter, your house is suddenly deprived of ventilation. That's when moisture, indoor air pollutants, and odors can really become troublesome.

A relatively new type of exhaust-only system, developed by American ALDES and DEC International, among others, addresses this problem nicely by using ducts and passive inlet vents.

The fans used in these central exhaust systems are much smaller and less powerful than whole-house fans and are designed to run virtually all the time. The intake side of the fan is fitted with a special manifold that has several ports or duct connectors. Lengths of flexible duct run from the manifold to air grilles located in the bath, kitchen, and utility room. Stale, moist air is gently pulled out of those rooms 24 hours a day and vented outside. If more spot ventilation is required (when the bathroom gets steamy, for example, or the kitchen smoky), the system is equipped with manual in-room controls that can temporarily boost the fan speed up a notch.

Instead of drawing fresh air through open windows (the way whole-house fans do), this system uses passive air inlets that are typically located in the living room and bedrooms. These through-the-wall inlets have dampers that react to the slightly negative pressure the fan exerts on the house, admitting clean, fresh air as needed, without creating a draft. You might call it "controlled infiltration."

This type of central exhaust system is best suited for newer, tightly built homes that need an affordable whole-house ventilation system. But it could also be a good choice for ventilating an older house that's been tightened up. Because the air flow on central exhaust systems is a lot less than whole-house fans, they aren't nearly as beneficial for summertime cooling. The installed cost for a central exhaust system starts at about $600, but there are significant savings in not having to install separate exhaust fans for the bath or kitchen.

A still more sophisticated approach to exhaust-only ventilation is to use an exhaust air heat pump. The system, invented by the Swedes and further developed in the U.S. by DEC International, features an air-to-water heat pump mounted atop an 80-gallon water heater.

Warm, stale air collected from the house is ducted through the heat pump before it's exhausted outside. The heat pump extracts heat from the exhaust air and dumps it into the water heater, cutting the home's annual water heating costs by 50 percent or more. The system can also be configured to provide supplemental space heating. Fresh air is introduced to the house through the same type of passive, through-the-wall inlets that I described above. Installed costs for an exhaust air heat pump will range from $1,600 to $2,500.

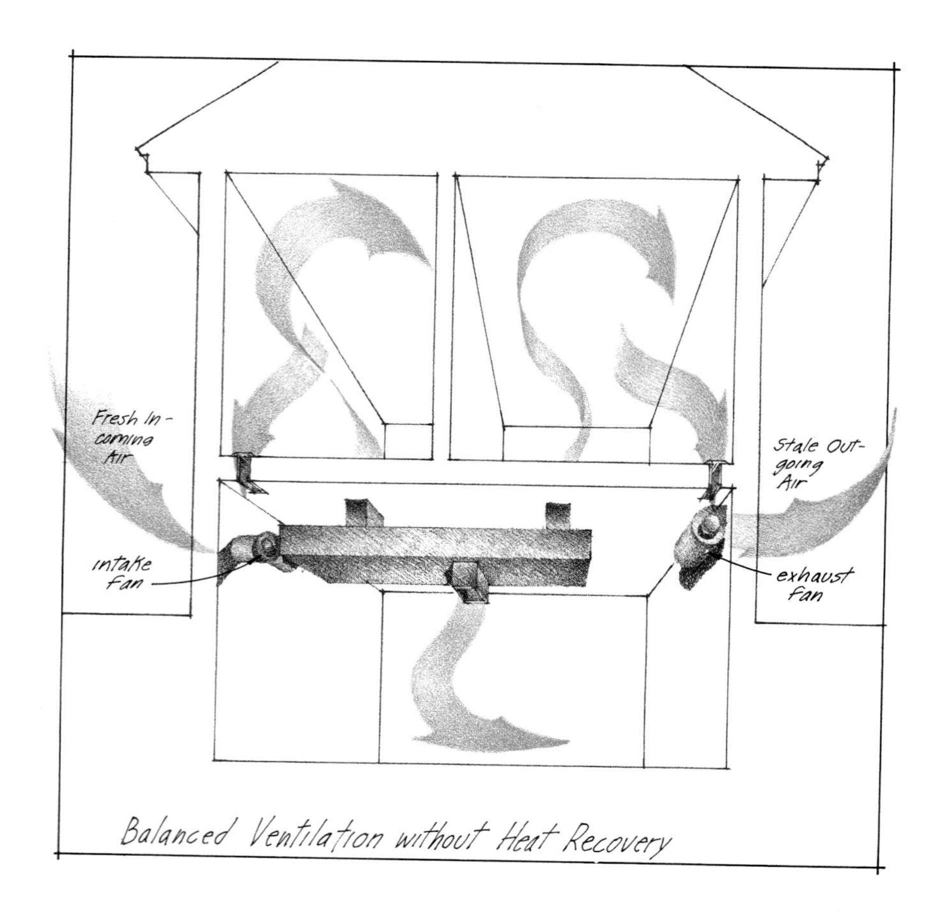

Balanced Ventilation without Heat Recovery

With any type of exhaust-only system, there are important safety tips to keep in mind.

Since exhaust-only ventilation systems suck air out of the house, they can cause a potentially dangerous depressurization around open-combustion appliances. Without a ready supply of combustion air, furnaces, boilers, gas water heaters, wood stoves, and fireplaces can backdraft lethal fumes into the house. So if you're having new ventilation equipment installed, make sure that it doesn't compromise the air supply to any open-combustion appliances.

Another concern is that depressurizing the house could draw radon — a cancer-causing soil gas — up through the basement or crawl space. With that in mind, I recommend that you test your house for radon *before* installing any new ventilation equipment. In fact, a radon test is a good idea regardless of your plans. Simple-to-use charcoal test canisters are available through state health offices and hardware stores for about $20.

Balanced Ventilation Systems

Balanced ventilation systems have two fans: one for drawing fresh air into the house, the other for exhausting stale air. In theory at least, the system is supposed to be balanced so that the air drawn into the house is equal to the air blown out, and the house is neither pressurized nor depressurized.

Balanced systems typically have ductwork connected to the intake fan so that fresh air can be effectively distributed to registers in various rooms. Since the incoming air would be uncomfortably cold in the winter, an in-duct electric heating element can be used to pre-warm it. A good-quality installation would have a complete ventilation loop, with return air registers and ducts terminating at the exhaust fan.

There are three drawbacks to this type of system that are serious enough to torpedo your comfort and bloat your utility bill. First, you have to heat the incoming air electrically in the winter. Second, the exhaust fan blows heated air out of the house in the winter, which increases the demand on your furnace. Third, if you use the ventilating system in the summer, the intake fan can pull hot humid air into the house, increasing the load on the air conditioner.

HVAC engineers have come up with a clever solution to these problems. It's called a heat recovery ventilator, or HRV.

As you can see from the schematic drawings on page 000, various types of HRVs share a common working principle: They transfer heat between two adjacent streams of air that are moving in opposite directions at different temperatures.

Indoor Air Quality

Though the terms "indoor air pollution" and "sick building" are fairly new, the problem itself is old as the hills. Picture an early family of *Homo sapiens*, huddling around the fire inside a poorly vented cave or wigwam. I won't take the time here to list all of the carcinogens and mutagens contained in wood smoke, but it's a pretty good bet that those prehistoric families had a lot more to fear from lung disease than they did from wild beasts.

Our modern consumer society has brought with it 101 new sources of indoor air pollution, from aerosols to asbestos to adhesives. Many of these pollutants were introduced into our homes unwittingly as the hidden by-products of certain building materials and consumer products. Others, like radon and plant pollen, are a natural consequence of where we live. Still others, like cigarette smoke, are a poison of our own choosing.

One of the most telling and highly publicized revelations about indoor air quality came from an experiment that the U.S. Environmental Protection Agency ran in the early 1980s. Three hundred and fifty homeowners in northern New Jersey were fitted with personal air monitors so that the quality of the air they breathed during a typical day could be measured wherever they went.

As you've probably guessed already, the most severely polluted air the homeowners encountered, in virtually every case, was *inside their own homes*. For the eleven target chemicals that were monitored in the experiment, including a number of carcinogens, the average indoor air showed concentrations that were two to five times worse than outdoors — and this in northern New Jersey, where the outdoor air isn't exactly pristine. In the worst cases monitored, the target chemicals were 10 to 100 times more concentrated indoors than out.

Says EPA scientist Lance Wallace, who oversaw the experiment: "The tests showed that the most pressing threat from air pollution isn't from the petrochemical or rubber plant out at the edge of town — it's from the paints and adhesives and other toxic products that we bring into our own homes."

While it may be tempting to blame indoor air-quality problems on energy conservation, researchers have found no reliable link between how tightly a house is built or retrofitted and problems with indoor air quality. To put it another way, a house that's outrageously leaky may have serious indoor air pollution while its tightly built neighbor has none.

"Tightening up a house with caulk and weatherstripping isn't what causes the problem," notes John Spears, of Geomet Technologies. "The problem lies at the source of the pollution."

With that in mind, the best way to address an air-quality problem is to *identify* and *reduce* the source of the pollution. If a pollutant can't be avoided or controlled at the source, ventilation and air filtration may provide alternative solutions.

Obviously, the first step in this process is to find out whether or not you have a problem. In many cases your eyes and nose and the way you feel will alert you. Chemicals that are unhealthy to breathe often signal their presence with an unpleasant odor. So do molds, fungi, and mildews. (Indoor air "fresheners" and room "deodorants" only mask the problem, and can sometimes be irritants in their own right.)

Watering eyes, sore throats, headaches, nasal irritations, upper respiratory problems, nausea, vomiting, and/or rashes can be telltale symptoms that you're suffering from indoor air pollution.

Sometimes the signals are more subtle. Maybe the house simply has a "stuffy" feel to it, without any discernable odors. Or it dawns on you that you always seem to feel better when you're outside.

Whatever the case, listen to what your body is trying to tell you. Ignoring the warning signs could lead to serious health problems.

Having said that, I should point out that some indoor contaminants — including radon, asbestos, and lead — can't be detected through the senses and may not produce early physical symptoms to warn you. The chart on pages 222–225 provides information on the different categories of pollutants. Your state health office, the U.S. Environmental Protection Agency, and other organizations can provide additional help. (See page 227.)

A second chart, on page 221, shows the effectiveness of different types of air filters. On the low end of the efficiency range is the dust lint filter that comes standard on many forced air furnaces. These inexpensive filters, made of woven fiberglass or steel mesh, are really designed to keep the heating and cooling coils and fan clean rather than to protect human health. Still, they're helpful in removing plant spores, pollen, and large-caliber dust particles from the air. If you have this type of filter, it's very important that you clean or replace it frequently, since a dirty filter can become a *source* of pollutants.

Electronic air cleaners represent the top-end choice for residential applications, removing up to 95 percent of indoor pollutants, including airborne bacteria and many viruses. But electronic air cleaners will add $500 to $700 to the cost of a forced air system and they require frequent maintenance.

Self-charging mechanical filters, which use static electricity to enhance their air-cleaning ability, and media filters, which are similar to the air filters used in cars, represent middle-range choices in terms of cost and efficiency.

Not included on the chart are room air cleaners, which can be effective in removing dust, cigarette smoke, and pollen. Room models are rated according to their Clean Air Delivery Rates, which are certified by the Association of Home Appliance Manufacturers (AHAM). (For a listing of certified manufacturers and a free pamphlet titled "Consumer Guide for Room Air Cleaners," write: AHAM, 20 North Wacker Drive, Chicago, IL 60606.)

Comparing Air Filter Efficiency

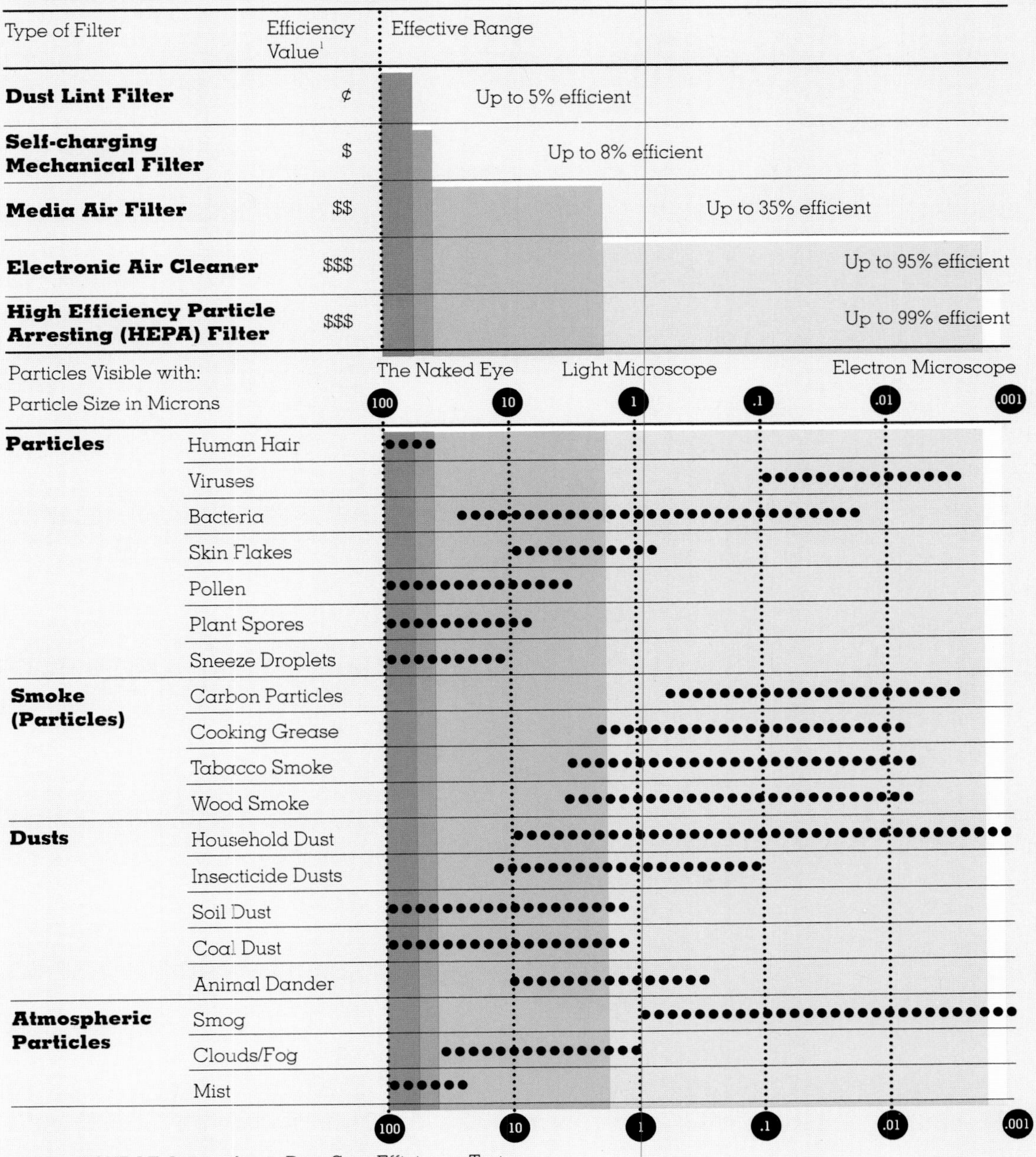

Type of Filter	Efficiency Value[1]	Effective Range
Dust Lint Filter	¢	Up to 5% efficient
Self-charging Mechanical Filter	$	Up to 8% efficient
Media Air Filter	$$	Up to 35% efficient
Electronic Air Cleaner	$$$	Up to 95% efficient
High Efficiency Particle Arresting (HEPA) Filter	$$$	Up to 99% efficient

Particles Visible with: The Naked Eye, Light Microscope, Electron Microscope

Particle Size in Microns: 100, 10, 1, .1, .01, .001

Particles	Human Hair
	Viruses
	Bacteria
	Skin Flakes
	Pollen
	Plant Spores
	Sneeze Droplets
Smoke (Particles)	Carbon Particles
	Cooking Grease
	Tabacco Smoke
	Wood Smoke
Dusts	Household Dust
	Insecticide Dusts
	Soil Dust
	Coal Dust
	Animal Dander
Atmospheric Particles	Smog
	Clouds/Fog
	Mist

100 10 1 .1 .01 .001

SOURCE: ASHRAE Atmospheric Dust Spot Efficiency Test

1. Efficiency value based on cleaning efficiency in relationship to cost

Some Significant Indoor Air Pollutants

Pollutant	Sources	Detection
Asbestos A natural mineral used in various building materials. If the fibers are released into the air and inhaled, they can be trapped in the lungs, causing disease.	• Some wall and ceiling insulation between 1930 and 1950. • Old insulation on heating pipes and old heating equipment. • Old wood stove door gaskets. • Drywall joint-finishing material and textured paint purchased before 1977. • Some older vinyl floor tiles. • Cement-asbestos millboard and exterior house wall shingles. (Note: If the asbestos fibers remain bound within these materials and do not create dust, the occupant is not at risk.) • Old fireproof cloth products. • Some sprayed and troweled ceiling finish plasters installed between 1945 and 1973.	• It can be detected by certified asbestos abatement contractors. • If you have an asbestos-containing product in the open that is disturbed by vibration or contact, asbestos particles may be in the air.
Cleaning Agents & Aerosols Most of these chemicals are hydrocarbons that act as solvents in cleaning agents. Also, hydrocarbons often serve as propellants for aerosol products.	• Cleaners, paints, hair sprays, glues, fabric softeners, pesticides, perfumes, deodorizers. Paint strippers containing methylene chloride are especially dangerous.	• Most of these chemicals have a detectable odor at high concentrations. • Low concentrations can be detected by professionals.
Combustion By-products Include particles, carbon monoxide (CO), carbon dioxide (CO_2), nitric oxide (NO), nitrogen dioxide (NO_2) and partially oxidized organics.	• Combustion, including gas ranges, wood stoves, tobacco smoke, unvented heaters, and fireplaces. • Automobile exhaust. • Incomplete combustion of fuels. • Unvented combustion gases.	• Combustion particles can usually be detected by their odor when at high concentrations. • The gases listed are odorless and colorless, but can be detected by professionals. • Certain combustion by-products such as tobacco smoke and wood smoke are particularly strong in odor.
Formaldehyde A colorless, water-soluble gas that may have a detectable odor at high concentrations.	• Some particle board, plywood, pressboard, paneling, carpeting, upholstery and furniture. • Urea formaldehyde insulation (used in walls, mainly during 1970s). • Some household cleaners and deodorizers. • Gas stoves, tobacco smoke and poorly vented wood stoves. • Mobile homes often have higher concentrations than wood-framed homes.	• Pungent odor from certain types of new carpeting or particle board. • Professionals can detect with special tests.

In the "All New This Old House" built in Brookline, Massachusetts, we installed an HRV with a counterflow heat exchanger that has worked with very satisfactory results. In the winter, when the ventilation system is exhausting stale *warm* air, the HRV recovers heat from that outgoing stream of air and uses it to pre-heat the cold incoming air. The system works in reverse in the summer,

Potential Health Effects	Control Methods
• Skin contact with asbestos may cause severe irritation. • Long-term inhalation can lead to coughing, chest pain, weakness, and lung cancer. • Mesothelioma (cancer of lining of chest and abdomen). (Lung cancer risk many times worse for smokers.) • Asbestosis (scarring of the lung tissue).	• Do not use materials containing asbestos. • Use fiberglass wood stove gaskets. • Do not disturb materials containing asbestos. • If a material containing asbestos is damaged or needs repair, contact a certified asbestos abatement contractor. • Do not sand siding, floor tiles, or other material manufactured with asbestos.
• Irritation of mucous membranes of nose. • Headaches. • Heartburn and abdominal pain. • Mental confusion. • Possible serious long-term effects.	• Read labels before using. • If you must use hazardous chemicals, try to use outside of your house. • If used inside, cross-ventilate area with fan and open windows. • Follow directions and do not mix products together. • Store products safely, cap tightly. • When possible, use products that are less hazardous. • Air out freshly dry-cleaned clothes before wearing.
• Impaired vision. • Headache. • Nausea. • Drowsiness. • Emphysema. • Heart disease. • Respiratory infections. • Reduced lung capacity. • Lung cancer. • Death from carbon monoxide.	• When cooking, use a kitchen exhaust fan that is vented to the outside. • Keep combustion appliances properly adjusted. • Do not use unvented combustion appliances. • Do not warm up car in attached garage. • Use pilotless ignition on gas appliances. • Exhaust smoking area and provide a fresh air supply. • Inspect and maintain all fuel-burning equipment. • Properly maintain chimneys, solid fuel stoves, and fireplaces to avoid leaks.
• Eye and skin irritation. • Feeling of pressure in head. • Dizziness. • Nausea. • Upper respiratory irritation. • Lower respiratory irritation and pulmonary effects. • Pulmonary edema, pneumonia. (Note: The sensitivity to formaldehyde varies widely.)	• Remove source if identifiable. • Try to avoid products which contain high levels of formaldehyde. • If plywood, particle board, or pressboard is used in the dweling, use low-formaldehyde types if possible. • Avoid high relative humidity within the dwelling. • Seal particle board, pressboard, and paneling containing high levels of formaldehyde with varnish or vinyl wallpaper. • Ventilate space containing formaldehyde odors. • Professional chemical fumigation may neutralize formaldehyde. (Note: It may be difficult to determine the level of formaldehyde in a product or whether product contains formaldehyde.)

using the cool, stale air that's flowing out to pre-cool the warm air coming in.

Most heat recovery ventilators are designed for whole-house ventilation, requiring a complete system of supply and return ducts. The good news is that an HRV can be readily added to an existing forced air heating system, taking advantage of the ductwork

Some Significant Indoor Air Pollutants (continued)

Pollutant	Sources	Detection
Organisms and Microbes Broad terms encompassing most microscopic particles, living plants and animals, including fungi, mold, mildew, house dust mites, virus, bacteria, animal dander, and respirable particles.	• Mold, mildew and other fungi thrive in damp, humid places. • Humidifiers. • Air conditioners. • Heating systems if dirty or wet. • House dust.	• Most are only detectable with special equipment used by professionals. • Odor of mold and mildew.
Radon A radioactive gas formed as a by-product in the decay chain of uranium 238. When radon is in the air, its progeny (radioactive off-spring) are also in the air.	• Radon gas emanates from rocks and enters through cracks and holes in the foundation, slab or dirt floor. • Water from some private drilled wells. • Some building materials, such as granite.	• Odorless and colorless. • Inexpensive charcoal canisters or track-etch detectors that are sent to a laboratory for analysis. • Special electronic meters which measure the radioactivity emitted during the decay of radon and its progeny. • Laboratory-conducted test of water from private drilled wells.
Water Vapor Water in gaseous form, an ever-present constituent of air. Too much water vapor in the air can be harmful while too little can be uncomfortable. Water vapor can act as a vehicle and catalyst for other pollutants.	• Household activities such as showering, bathing, and cooking. • Human respiration. • Unvented combustion. • Improper drainage around house.	• Excessive condensation on windows. • Condensation on walls and ceilings. • Musty smell. • Can be tested with a sling psychrometer.
Lead Highly toxic heavy metal.	• Paint made before 1978. • Lead from solder or pipes can leach into water.	• Do-it-yourself test kits available for paint, water, and pottery. • Professional testing.

SOURCE: R.J. Karg Associates, Housing Resource Center

that's already there. The bad news is that retrofitting a house that doesn't have ductwork can be quite a bit more difficult and expensive. In either case, the key to a good installation is an able contractor.

Some companies are manufacturing ductless HRVs that can be simply mounted through the wall or ceiling. These relatively inexpensive units are only intended to provide heat-recovery ventilation to the immediate room, though adjoining rooms would certainly benefit if the connecting doors were left open.

Like the fans I talked about earlier, HRVs are rated by the

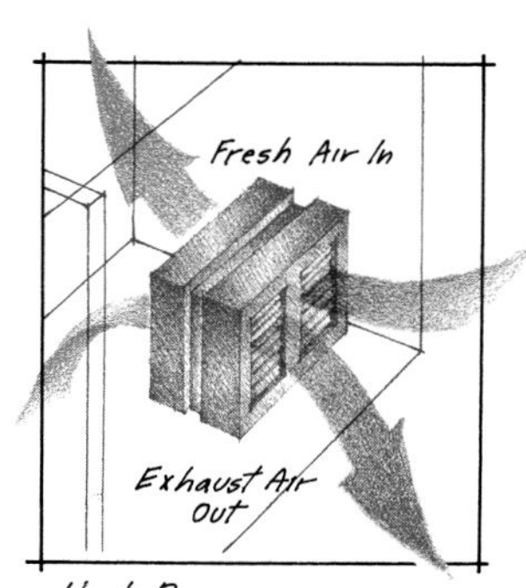

Heat Recovery Ventilation for through the Wall Installation (Summer Mode)

Potential Health Effects	Control Methods
• Allergic reactions. • Headaches. • Flu symptoms.	• Maintain dwelling relative humidity between 30% and 40% in winter. • Avoid using carpets in bathrooms, laundry room, and other high-moisture areas. • Wash inside of humidifier weekly with mild chlorine solution. • Use exhaust fans that vent to the outside in bathrooms and kitchens. • Clean showers and other mildew-prone areas frequently. • Control source of irritants. • Ventilate crawl spaces and basements during warm weather. • Install polyethylene ground cover over bare ground in crawl space and basement.
• Major cause of lung cancer. Risk many times greater for smokers.	• Seal cracks and holes in foundation walls and concrete slabs. • Install water traps in basement floor drain pipes. • Ventilate basement or crawl space by exhausting air or by installing heat recovery ventilation which exhausts and supplies air. • Tightly seal sump holes in basement slabs, or, in special cases, ventilate sump holes to the outside. • Install 6 mil polyethylene under new basement slabs. • Install a ventilation system under basement slabs. • If well water contains radon, filter water with special charcoal filter or aerate water with appropriate system. • Retest for radon after implementing control methods.
• Increase in allergic reactions, including asthmatic attacks, to house-dust mites, mold and mildew, and chemicals. • If formaldehyde is present, its concentration increases at higher relative humidities.	• Control water vapor production within dwelling. • Keep relative humidity between 30% and 40% in winter. • Do not vent clothes dryer to the inside, especially a gas dryer. • Use fans that vent to the outside in bathrooms and kitchens when rooms are in use. • Try to avoid using a humidifier. • Drain water away from foundation. • Cover dirt floors in crawl spaces and basements with plastic sheeting.
• Even low-level exposure can cause neurological, kidney, blood, and reproductive disorders and hyptertension. • Children and pregnant women especially susceptible.	• Replace or cover painted surfaces. • Use nontoxic liquid paint stripper or low-temperature heat gun. Do *not* use torch or high-tempertture heat gun to strip paint. Do *not* sand or blast lead paint off without taking adequate precautions. • Let water run for several minutes each morning to flush lines. • Replace old lead pipes. • Never use lead solder.

amount of air they can handle. Whole-house systems have CFM capacities up to 600, with total installed costs ranging from $900 to $3,000. Through-the-wall models are typically rated under 100 CFM and cost $400 to $600 installed.

HRV technology has made gigantic strides in the past ten years, with some models approaching 85 percent efficiency. That means that they can recover 85 percent of the sensible heat in an outbound stream of stale air and put it right back into the house. (They are also 85 percent efficient in pre-cooling incoming air in the summer.)

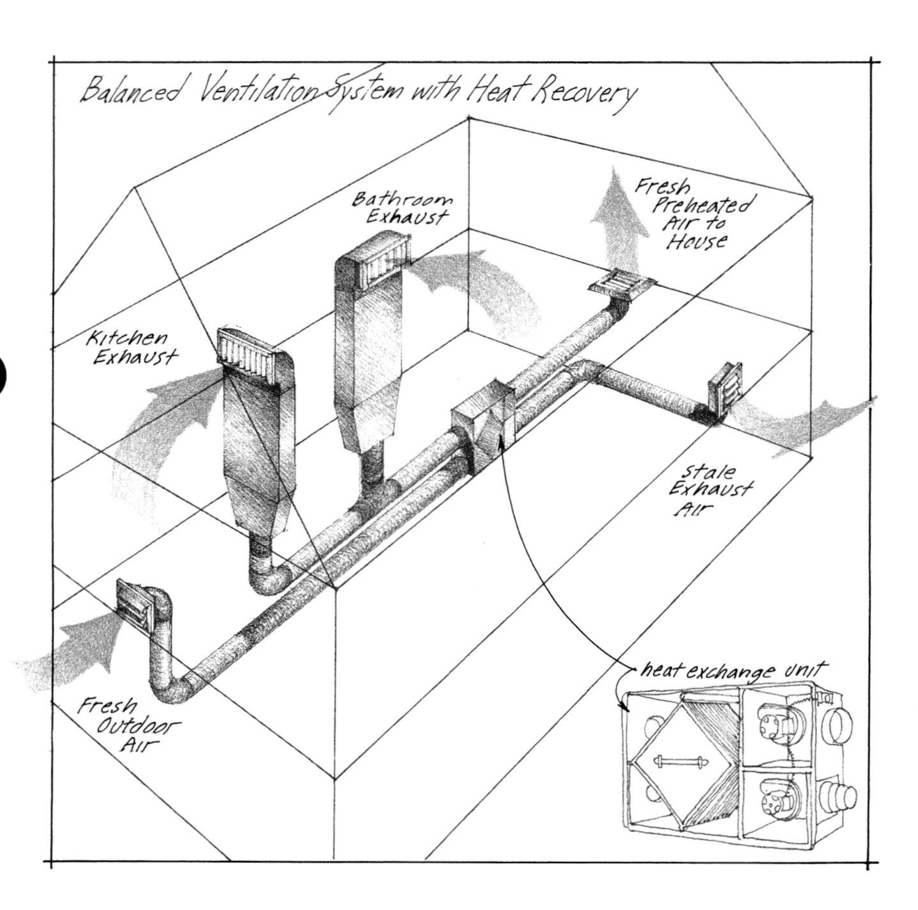

Some HRVs, like the ones sold by Honeywell and NewAire, can also transfer water vapor from one airstream to another. This enables them to humidify incoming air in the winter and dehumidify it in the summer, a powerful advantage in improving comfort and reducing the load on your heating and cooling equipment.

Heat recovery ventilators have made the biggest impact in the northern tier of the U.S., where they've become a popular choice for tightly built homes, and in Canada, where they're a required component in all new houses. While early technology and marketing efforts have focused on cold climates, researchers are now suggesting that HRVs can make good economic sense even in southern climates that are dominated by air conditioning.

Most of the problems with HRVs, as revealed in a recent Canadian study, aren't due so much to equipment failure as they are to faulty installation work.

"Most of the problems we've encountered result directly from poor or incomplete installations, carried out by unqualified tradespeople or contractors who don't have a clear understanding of ventilation principles or haven't kept abreast of technical ad-

For More Information

U.S. Environmental Protection Agency
Public Information Center
401 M Street SW
Washington, DC 20460

- "The Inside Story: A Guide to Indoor Air Quality"
- "Indoor Air Facts: Residential Air Cleaners"
- "Residential Air-Cleaning Devices"
- "Use and Care of Home Humidifiers"
- "A Citizen's Guide to Radon: What It Is and What to do About It
- "Radon Reduction Methods: A Homeowner's Guide

American Lung Association
1740 Broadway
New York, NY 10019

- "Biological Pollutants in Your Home"
- "An Update on Formaldehyde"
- "Combustion Appliances and Indoor Air Pollution"
- "Asbestos in Your Home"

Home Ventilating Institute
30 W. University Drive
Arlington Heights, IL 60004

- "Home Ventilating Guide"
- "Certified Products Directory"

Association of Home Appliance Manufacturers
20 North Wacker Drive
Chicago, IL 60606

- "Consumers Guide for Room Air Cleaners"
- "Product Directory: Humidifiers"

Consumer Product Safety Commission
Washington, DC 20207

- Fact sheet on humidifiers

Air Conditioning & Refrigeration Institute
Dept. U-181
P.O. Box 37700
Washington, DC 20013

- "Indoor Air Quality Briefing Paper"
- "Breathing Clean: How Air Filters Provide Cleaner Living"
- "Air Conditioning and Refrigeration Equipment: General Maintenance Guidelines for Improving Indoor Air Environment"

vances," says engineer William Mayhew, who oversaw the Canadian study.

That underscores the point I've been making all along: *Buy the contractor first, then the hardware.*

A good additional source of information on ventilation is *The Home Ventilating Guide*, available through the Home Ventilating Institute, 30 West University Drive, Arlington Heights, IL 60004.

●

Air Conditioning

If the passive cooling and ventilation strategies outlined in the last chapter aren't enough to keep you comfortable in the summer, you'll want to add a more powerful cooling system.

In New England, and the eastern part of the United States in general, that means buying an air conditioner with a compressor and some type of pressurized refrigerant.

But first, I'd like to talk a little about evaporative cooling, a truly elegant and ancient technology. In fact, archaeological records show that an Assyrian merchant — who lived 3,000 years before Christ — had already discovered the advantages of evaporative cooling. During spells of hot weather, he'd order his servants to spray water over the walls and floors of a room located directly underneath his courtyard. Though the Assyrian's system needed plenty of cheap labor to work, and no patent was ever issued, he probably deserves credit as the inventor of evaporative cooling.

Unlike refrigerant air conditioners, which are complex and expensive, modern evaporative coolers — or "swamp coolers" as they're sometimes called — are relatively simple and inexpensive machines. Imagine, if you will, a curtain of wet burlap hung over a window with the wind blowing through and you've pretty much got the idea.

Instead of burlap, evaporative coolers use spongelike pads that are continually wetted by a circulating pump. An electric blower pulls air through the pads, where it's chilled and filtered, and passed into the home directly or through ductwork. By this means, evaporative coolers take advantage of the heat lost through the evaporation of water to provide low-cost cooling to the home.

The installed cost of an evaporative cooler will run from $600 to $800 for a window unit and from $1,200 to $1,400 for a fully ducted system, which is quite a bit less than air conditioning. But more importantly, evaporative coolers cost only one-fourth to one-half as much to operate. When you add it all up, the life-cycle savings versus refrigerant air conditioning can be as much as $5,000!

But cost isn't the only advantage of evaporative coolers. Unlike most other air conditioning systems, which recirculate indoor air, evaporative coolers introduce large volumes of fresh, clean air into the home, which helps ensure good indoor air quality. Moreover, they don't use R-22 or other refrigerants, which have been pegged as a threat to the earth's ozone layer.

Unfortunately, evaporative coolers don't work very well in humid areas. They perform best in the hot, dry regions of the West — places like Los Angeles, Phoenix, Salt Lake City, and Dallas.

Good as they are, evaporative coolers can't match the carefully controlled cooling power and dehumidification offered by a compressor-driven air conditioner or heat pump.

It's ironic, I think, that when Dr. Willis Carrier invented refrigerant air conditioning back in 1902, he wasn't sure just how much

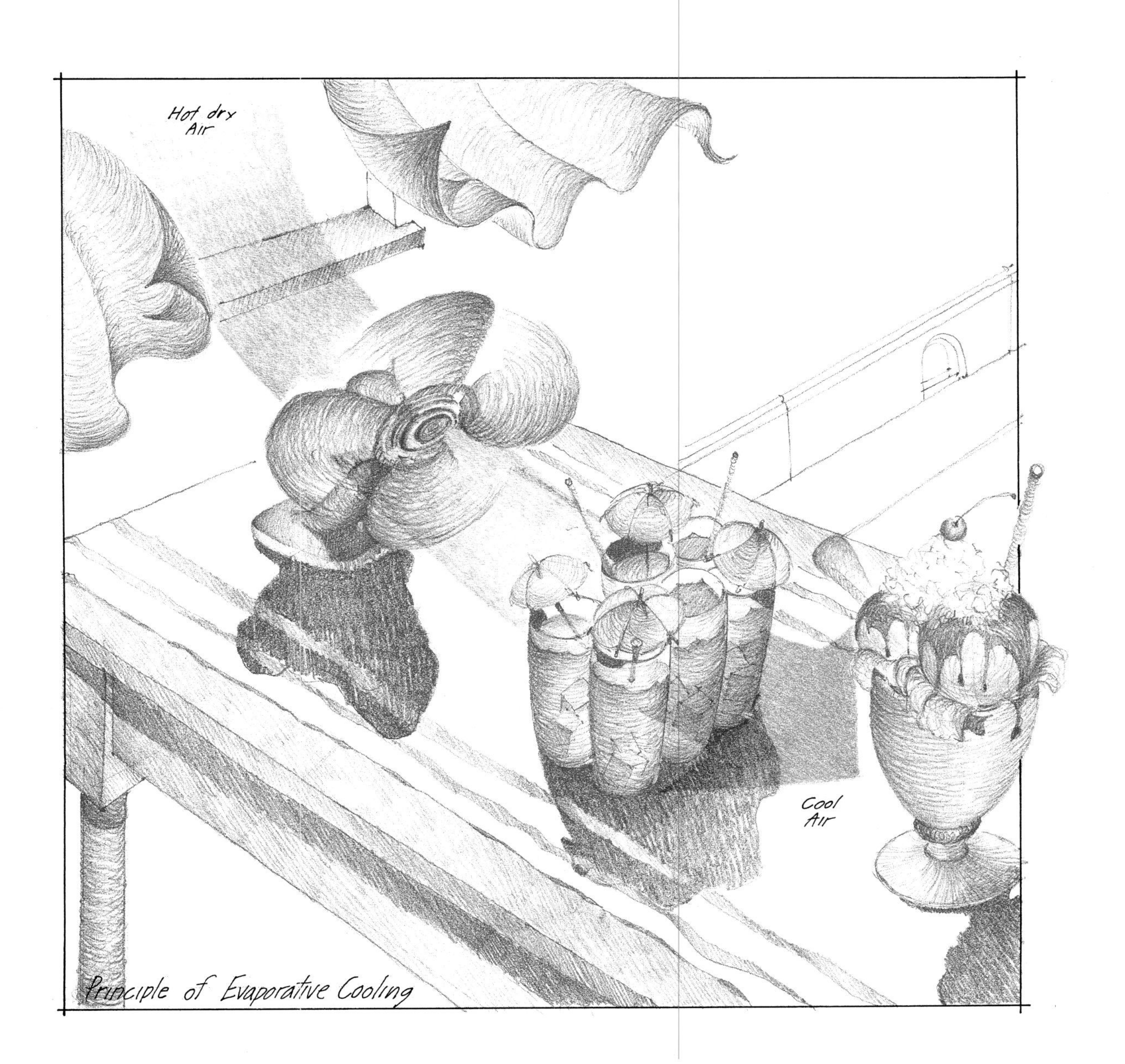

Principle of Evaporative Cooling

interest there was going to be — so he staged a boxing match to attract a crowd and used that somewhat rowdy occasion to show off his new invention. Ninety-odd years later, no one remembers who the fighters were that night. But find me, if you can, someone who hasn't heard of air conditioning!

Factories, businesses, and theaters were the first to exploit Carrier's new technology. It wasn't until 1914 that the first house — the Charles Gates mansion in Minneapolis — was equipped with mechanical air conditioning. Even so, residential air conditioning remained a novelty for the rich and famous until 1952. That's when a builder in St. Louis decided to take a chance and offer air conditioning in his new housing development. Hundreds of curious people showed up to see this wonder of wonders: a medium-priced home with air conditioning! The entire subdivision was sold out in two weeks.

Today, more than 27 million homeowners enjoy central air conditioning and another 17 million have one or more room air conditioners. Even more impressive is the fact that 75 percent of all the new houses built in the United States now come equipped with central air.

How It Works

As you can see from the diagram on page 231, central air conditioners use a compressor, two coils, and a chemical refrigerant to extract heat from the indoor air and transfer it outside. As the liquid refrigerant moves through the indoor coil, or evaporator, it absorbs heat from the airstream and changes into a gas. The gas is then routed through the compressor, which compresses the gas and raises its temperature. As the gas flows through the outdoor coil, or condenser, it gives up its heat to the outside air and condenses back into a liquid.

Though most people don't realize it, air conditioning actually cools you in two ways:

First, it handles the *sensible heat load,* which refers to the need to remove heat energy from the air and thereby lower its temperature. Second, it handles the *latent cooling load,* which refers to the need to remove water vapor from the air.

In addition to cooling and dehumidification, a good central air conditioning system will also control air circulation and ventilation, and provide a means to clean the air.

Heat pumps, which burst upon the scene in the early 1960s, are a special type of air conditioner that can be reversed in the winter to provide heat. During the summer, the heat pump's compressor and refrigerant loop function very much like a standard air

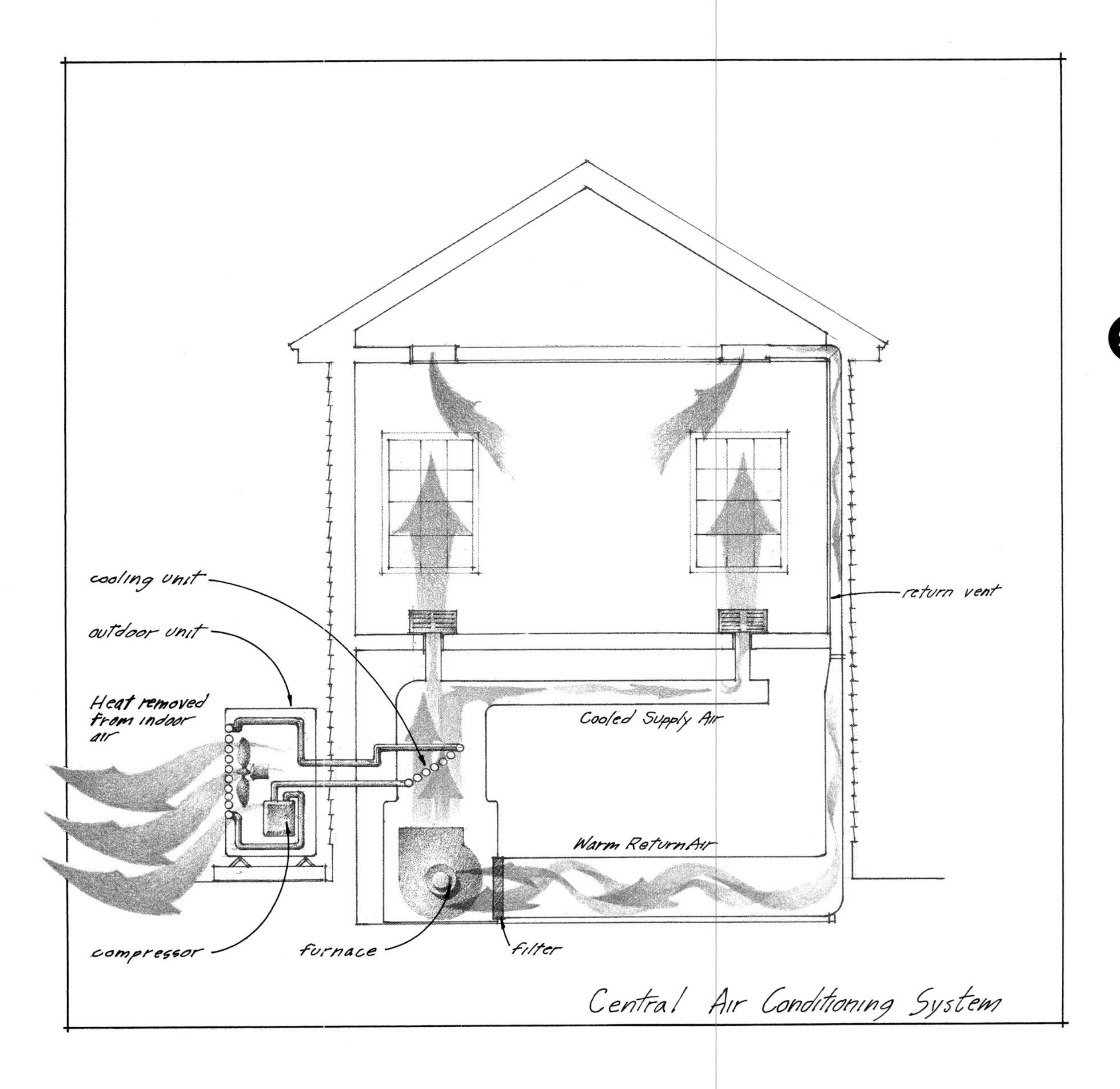

conditioner, drawing heat out of the indoor air and expelling it through the outdoor coil. But in the winter, the process is reversed — the heat pump absorbs heat from outdoors and pumps it inside.

A lot of people scratch their heads when they hear that, and ask: "How can a machine extract heat out of cold air?" But "cold," it turns out, is a very relative term. At 0°F, the air still contains more than 80 percent of the heat it held at 100°F. It's that large portion of remaining heat that the heat pump taps and transfers indoors.

Nevertheless, a heat pump's heating efficiency *is* affected by falling temperatures. As the mercury drops toward freezing and the heat pump loses some of its efficiency, an electric resistance heater built into the air handler automatically kicks in to make up

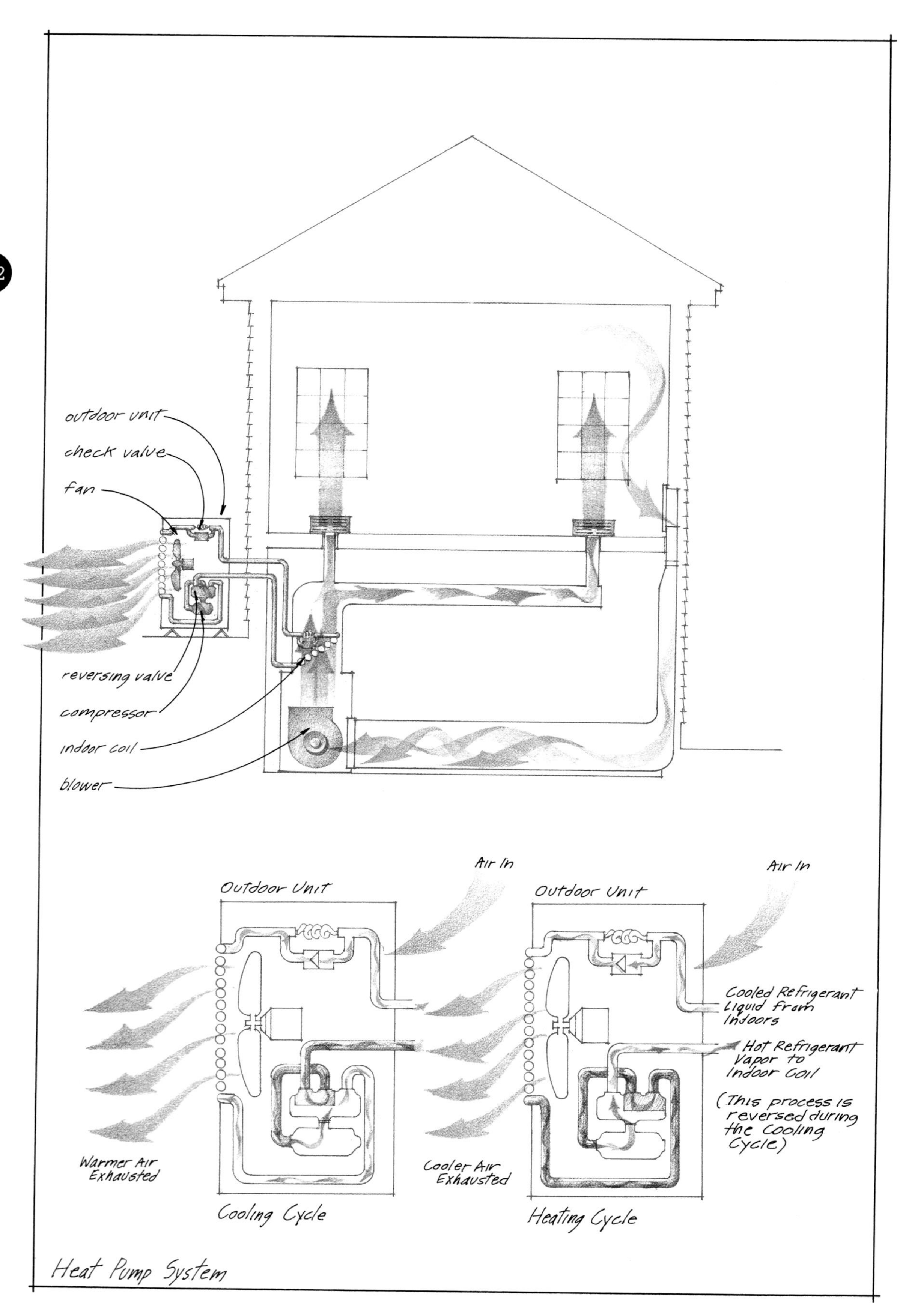

Heat Pump System

Trethewey's Tips: Taming Runaway Heat Pumps

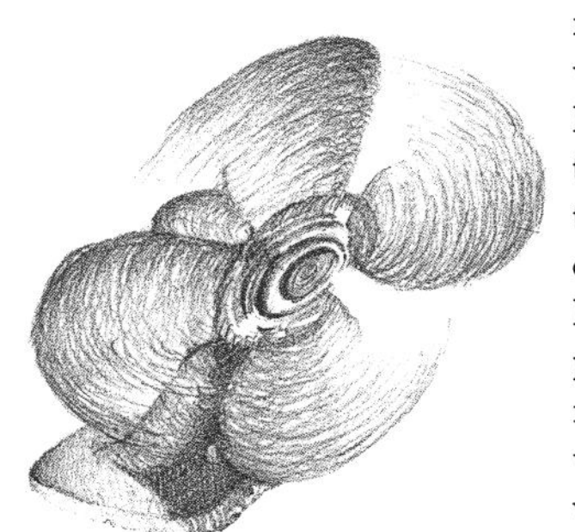

If you own an air-to-air heat pump that's burdening your family with high electric bills, a change in the way you control the system could save you a lot of money.

By installing a new thermostat that has a gradual ramp-up feature, the heat pump can cycle off and on without resorting to its backup resistance heater so often. It's that electric resistance heater, designed to cut in when there's a sudden demand for heat or when the outdoor air temperature drops below a certain threshold, that really chews up the electricity — and your hard-earned money! In selecting a new thermostat for your heat pump, make sure that you buy a model that's easy to operate. Otherwise, you won't use it to your full advantage.

Another way to tame a runaway heat pump is to install a thermostat cut-out switch that disconnects the strip heater altogether when the outdoor temperature climbs above 40°F. Above that temperature, the heat pump should be able to warm the house comfortably without using the backup element. In a test program conducted in Auburn, California, homeowners enjoyed a typical savings of about 8 percent on their heat pumps after cut-out switches were installed.

Of course, the strip heater isn't the only possible source of discomfort and high cost. Leaking ducts, dirty coils and filters, and too little or too much refrigerant are also common culprits.

the difference. That, of course, is when things start to get expensive.

The "This Old House" crew has installed air-to-air heat pumps on several different projects. Three of these were in the deep South — on houses in Tampa, New Orleans, and Phoenix — where the air conditioning load was our main concern and where air-to-air heat pumps can really shine.

Only twice have we used them in cold New England. The first time, as I recall, was in the Bigelow house, in Newton, Massachusetts, back in 1981. The owners were keen on having central air conditioning, especially in the property's outlying buildings, which were to be remodeled and sold as condos. They also liked the idea of providing heating and cooling to each unit with a single, compact machine. Though an air-to-air heat pump wasn't my first choice then, and probably wouldn't be today, that's what the owners wanted, so that's what we gave them.

The only other New England project in which we used an air-to-air heat pump was in the old house in Wayland, Massachusetts. Since the owners there wanted to add air conditioning, we went ahead and installed a heat pump. But with an interesting twist.

To get around the cold-weather penalty, we used a dual heating system with weather-responsive controls. This strategy enables a homeowner to use the heat pump for heating so long as the outdoor temperature is 30°F or more. If the temperature dips below that threshold, where the heat pump's efficiency starts to fall, the controls automatically switch the heating load over to a more efficient source — in this case, a boiler.

Some manufacturers sell this type of "hybrid" or "dual-fuel" system — heat pump plus furnace — already packaged with the appropriate controls.

Though manufacturers have made steady improvements in both the heating and cooling efficiency of air-to-air heat pumps, I still have reservations about them as a sole source of heating in cold-weather climates. They are more appropriate, I think, in climates where there's more air conditioning load than heating, and in places where homeowners have no access to natural gas or other relatively low-cost heating fuels. For Yankees and other northern tribes, a better choice would be to use the dual-fuel approach I described above or, better still, a geothermal heat pump.

Tapping Mother Earth

Over the last decade or so, a new type of heat pump has entered the market that can deliver excellent heating and air conditioning efficiency *regardless* of the outdoor air temperature. These so-called geothermal heat pumps are of two general types: open loop and closed.

The open-loop variety, or "water source heat pump" as it's sometimes called, must have a water well, pond, river, lake, or ocean nearby to operate. As you can see from the drawings, a typical open-loop system has a water supply pipe on one side and a discharge pipe on the other. The heat pump, located inside the house, is equipped with a heat exchanger that draws heat out of the water in the winter and dumps heat into the water during the summer.

The reason this type of system is 20 to 30 percent more efficient than an air-to-air heat pump is because the geothermal properties of the earth keep groundwater hovering between 40°F and 60°F regardless of the season, while air temperatures swing wildly from subzero up into the 90s. (The use of lake or river water as a source for geothermal heat pumps is more restricted — usually to the South — since the water temperature from those sources usually isn't as constant as groundwater.)

Unfortunately, open-loop systems have some serious drawbacks. First of all, most homeowners don't have access to a suitable water well, deep pond, or river. And drilling a new well, as we've demonstrated on the show, can cost $5 to $20 a foot. Second, local, state, and federal laws controlling the use and discharge of groundwater are already strict, and getting stricter. Third, the presence of salt and other minerals in the water can clog and corrode the heat pump if it isn't properly built and maintained.

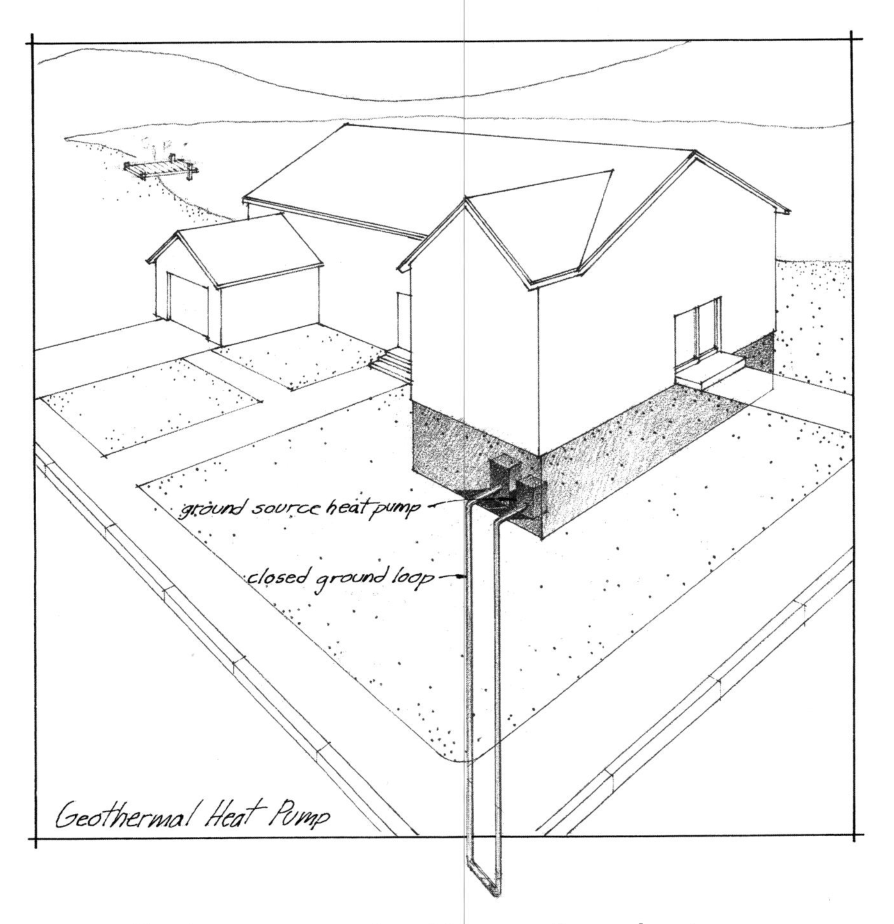

For these reasons, closed-loop geothermal systems, or "ground-coupled heat pumps" as they're sometimes called, are usually more practical. As shown in the drawings, this type of geothermal system relies on a continuous loop of polyethylene or polybutylene pipe that's buried in the earth. The configuration of the pipe and the depth at which it's laid depend on the size of the home's heating and cooling load and the geological conditions at the site.

A heat transfer fluid — usually water mixed with antifreeze — is circulated through the loop, up into the heat pump's heat exchanger, and back down into the loop again. In the summertime, heat is drawn out of the house and transferred into the ground; during the winter, heat is drawn from the ground and transferred to the house. No matter how hot or cold the air temperature outside gets, the earth — down below the frost line where those pipes are nestled — stays at around 50°F.

I think closed-loop geothermal heat pumps have a bright future across the nation, especially in new construction. Their main drawback is their relatively high first cost — say $6,000 to $8,000 for a three-ton system. It's the extra cost of trenching and installing the in-ground pipe loop that makes these systems generally more expensive, though excavation costs can vary enormously from one site to the next.

While a geothermal heat pump will cost more up front than other heating and cooling systems, it more than compensates with lower operating costs. At the Walden Pond subdivision in Carmel, Indiana — where all 126 homes have been equipped with geothermal heat pumps — the owners spend about $400 *less* each year than they would have if gas heat and conventional central air conditioning had been installed. So that extra up-front cost is quickly paid down by lower utility bills. (In some areas, utilities interested in reducing their peak load will actually help pay for a geothermal installation.)

Some of the leading makers of geothermal heat pumps include Water-Furnace, Florida Heat Pump, and ClimateMaster.

Another technology that holds great promise is the natural gas–fired heat pump, which is only now being introduced in the United States. York International, which worked with the Gas Research Institute to develop the technology, says that its new gas-fired heat pump can deliver air outlet temperatures that are 10° to 15°F warmer than electric air-to-air heat pumps and can operate more efficiently at lower outdoor temperatures. Several Japanese companies — including Yanmar, Aisin Seiki, and Yamaha — are also developing gas heat pumps.

Maintenance

Before you even consider buying new ventilation or air conditioning gear, make sure that what you've already got is properly maintained. As I explained in chapter 6, many of the comfort and cost problems related to forced air systems are actually due to poor maintenance, broken thermostats, duct failure, and weaknesses in the home's envelope that can be remedied with insulation, caulking, and weatherstripping.

Pacific Gas & Electric Co., the nation's largest utility, got to the heart of the matter when it launched its "Appliance Doctor" program in California a few years ago. Many of the homeowners who participated in that program had filed complaints with PG&E about high bills and gross discomfort. "I hate my heat pump" was one of the most frequent sentiments expressed among the complaining customers.

But when PG&E troubleshooters investigated the homes, they found that the most common source of problems were things like leaking ducts, dirty coils and air filters, too much or too little refrigerant in the system, and thermostat problems that caused the heat pump's electric backup heater to run too often or not at all.

PG&E found that a $400 package of maintenance work and simple repairs, which included repairing the ductwork, changing the air filter, cleaning the coils, and adjusting the refrigerant charge, would net the typical homeowner $1,950 in life-cycle savings, not to mention the substantial and immediate improvements realized in comfort. (The same improvements would net PG&E almost $1,200 in life-cycle savings, because it's cheaper for the utility to save energy than to build new generating plants.)

"In virtually every house we looked at, cooling energy use could be lowered by 10 to 30 percent without much effort," says John Proctor, who worked on the project.

One of the scariest findings from the PG&E program was that many of the homeowners involved had actually paid for a "professional" service call during the previous year, but were left with their comfort and cost problems mostly unresolved. The only explanation is that the service people who visited those homes before PG&E got there were either incompetent or dishonest, or both. Small wonder that so many utilities, institutions, and trade associations are calling for better HVAC training.

The chart I've assembled on page 238 outlines many of the do-it-yourself and professional maintenance routines that are needed to keep an air conditioning system healthy and happy.

I think one of the trickiest judgment calls for a homeowner (or contractor) to make is what to do if the heart of the system — the compressor — suddenly dies. If the compressor is still under warranty or the air conditioning system is relatively new — say, six or seven years old — I'd probably vote for replacing the compressor. But if the equipment is much older than that, you should consider buying a whole new system, since there's not that much life left in the old equipment anyway and a new system will save you money with its higher efficiency.

Sometimes, believe it or not, it can make good economic sense to junk old air conditioning equipment even if it's in running order. A test project in Oklahoma, conducted by the Alliance to Save Energy, U.S. Department of Energy, Oak Ridge National Laboratory, and Public Service Company of Oklahoma, found that when homeowners replaced their old, inefficient window air conditioners with higher efficiency models, their air conditioning bills fell by one-third or more — enough to pay for the new equipment in a few short years.

Air Conditioning an Old House

It goes without saying that a lot of houses in New England and other parts of the country were built before residential air conditioning

What the Doctor Ordered

Recommended Maintenance for Air Conditioners	Room	Central	Heat Pump	Can I Do It Myself?
Clean or replace air filters	X	X	X	Yes [1]
Clean the indoor coil with a vacuum or brush	X	X	X	Yes [2]
Clean outdoor coil with a vacuum or garden hose		X	X	Yes
Keep outdoor unit free of tall grass, shrubs, and other obstructions		X	X	Yes
Clean the fan blades	X	X	X	Yes [3]
Check fan belts for snugness and wear	X	X	X	Yes
Lubricate fan motors	X	X	X	Yes [4]
Cover or remove the unit for the winter	X			Yes
Seal the gaps between the air conditioner and the adjoining wall or window	X			Yes
Unclog the drain channel	X	X	X	Yes [5]
Test and tune the system; recharge it with refrigerant if needed	X	X	X	No [6, 7]
Check for leaking refrigerant	X	X	X	No
Test air flow	X	X	X	No
Make sure registers are properly oriented and free of obstructions	X	X	X	Yes
Vacuum the ducts		X	X	Yes [8]
Check for duct leaks and make needed repairs		X	X	No [9]
Check, repair, and/or improve duct insulation		X	X	Yes
Clean the humidifier		X	X	Yes [10]
Calibrate the thermostat	X	X	X	No [11]
Provide shade for outdoor unit		X	X	Yes [12]

See notes on opposite page.

became a practical reality. So the question comes up often: How do I add air conditioning to my old house?

The answer can be fairly straightforward if you already have forced air heating, since the supply and return ducts needed for central air conditioning are already in place.

Notes from chart, p. 238

1. Change or wash the filter in early spring, then once a month during the cooling season.
2. If it's accessible.
3. Scrape the blades clean with a screwdriver or putty knife.
4. Follow manufacturer's instructions. Some bearings may be sealed and self-lubricating.
5. Use a stiff wire to unclog the drain.
6. Room and central air conditioners should be professionally serviced every 2 to 3 years. Heat pumps, which run year-round, need to be serviced annually.
7. Refrigerants need to be handled in an environmentally safe manner.
8. Clean as far back into the registers as the vacuum will reach. For a more thorough job, you can hire a pro. But avoid firms that use gluelike materials to "seal" the dust inside the ducts. The sealers may contain formaldehyde or other harmful air pollutants.
9. Except for patching failed duct tape, the work is best left to a professional.
10. Some forced air systems have built-in humidifiers in the air handler or ductwork.
11. With an accurate thermometer and some patience you can detect a malfunctioning thermostat. But let your service rep recalibrate it or replace it.
12. Shading the outdoor unit with trees, shrubs, fencing, or a lean-to will increase its summertime performance.

Joe Iorio, president of Atlantic Heating and Air Conditioning, and a frequent participant in "This Old House" projects, estimates that 95 percent of the forced air heating systems in New England can be retrofitted with central air conditioning without heavy alterations to the ductwork.

"When there is a problem," says Joe, "it's usually a shortage of return air, which can be addressed by altering fan speeds."

If you do decide to add air conditioning to your forced air system, it becomes doubly important that the ductwork be tightly sealed and insulated. Otherwise you'll need a new rain hat (because of all the condensation that will come dripping off the ducts) and a very wealthy friend (to help you pay the utility bills).

But what about houses that don't have forced air heating?

For those homeowners, my first suggestion would be to consider an "oasis" approach to cooling. Instead of trying to retrofit the whole house with central air conditioning, which can be a very disruptive and expensive project, you choose two or three key areas — the master bedroom, for example, and the family room — and equip those alone with air conditioning. When you need shelter from the summer heat, you just plop yourself down in one of these oases for a while. You might also call this the "Let-the-Kids-Sweat" approach to cooling, since their bedrooms usually aren't air conditioned. (Don't feel guilty about this — their rugged little bodies can take the heat a lot better than yours!)

The most common way of cooling a single zone is with a room air conditioner, mounted in a window or through a hole cut in the wall. More than a dozen companies manufacture room air conditioners, with capacities ranging from 5,400 BTUs for a small room to 18,500 for a whopper. Energy efficiency ratings (EERs) range from about 8 to 13. If you expect to use the machine a lot and/or have high electric rates, it will pay you to invest in a high-efficiency model.

But room air conditioners have some fairly serious drawbacks. First of all, they're not very pretty to look at, either inside or out, and can be pretty noisy to boot. Second, if the unit is mounted in a window, it blocks the view and precludes using the window for natural ventilation. Third, room air conditioners are a notorious source of cold air leaks during the winter. Even if the seam around the unit is sealed tightly to the wall or window, cold air can still infiltrate through the inside of the machine. Fixing a plastic or canvas tarp over the outside grill can help a little. A better solution, if you can muster the energy, is to take the unit out during the winter and put it back in again in the spring.

The Japanese have addressed many of these problems with their "split-system" or "ductless" air conditioners. The heart of the

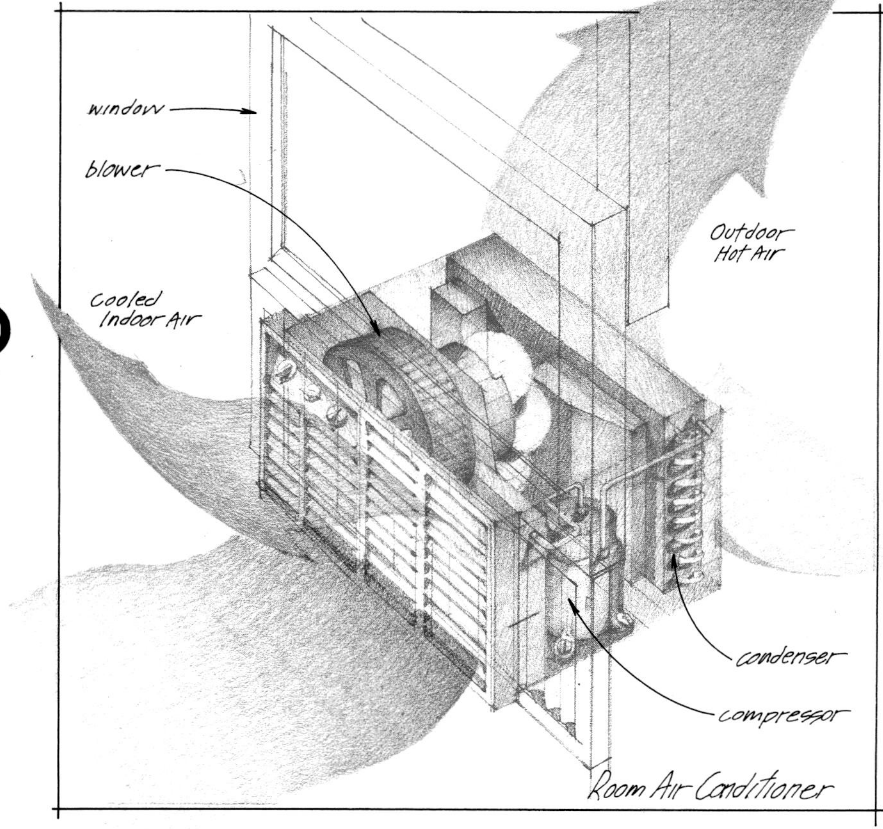

system is a compact outdoor unit, containing one or more compressors, which serve a matching number of indoor air handlers. As shown in the illustration, a special rigid conduit (2 to 3 inches in diameter) is run from the outdoor unit, through the wall of the house, to an air handler mounted inside. The insulated conduit contains a power cable, liquid and suction tubes for the refrigerant, and a condensate drain.

We first showed this type of system on television when we remodeled Mary Van and Jim Sinek's old house in Lexington, Massachusetts. As you may remember, Mary and Jim wanted to convert part of their house to a bed and breakfast, and felt that their summertime guests would really appreciate air conditioning.

We quickly rejected the idea of retrofitting the house with metal ductwork — it would have involved major surgery and might have bankrupted the owners in the process. None of us liked the idea of putting in room air conditioners either — it would have spoiled the looks of the grand old place.

I think Mary and Jim really appreciated the final advantages we achieved with a split system. Since the outdoor unit can be situated up to 50 feet away from the farthest room it serves, we were

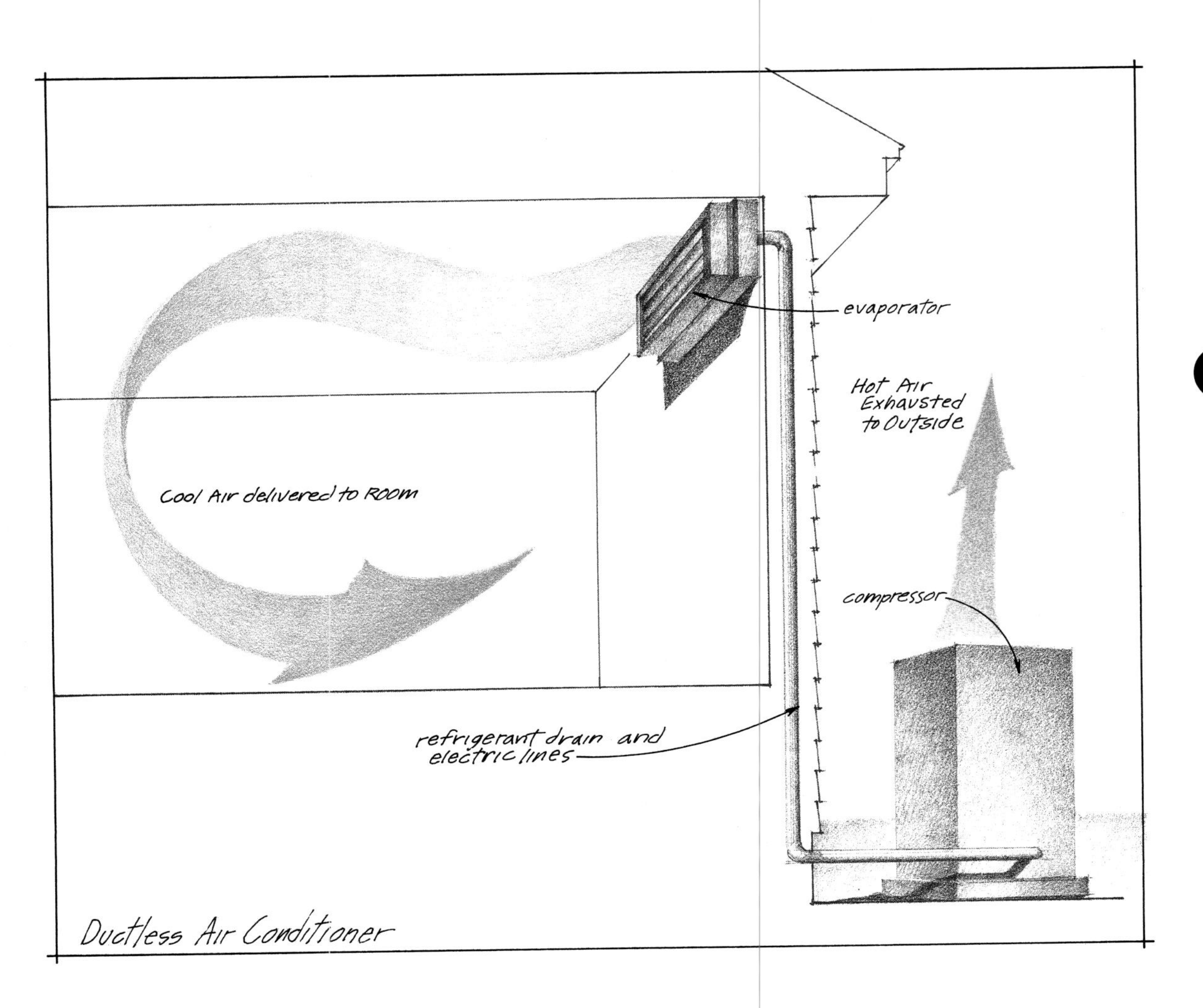

able to tuck it out back where it wouldn't "uglify" the house. That also meant that the noisiest part of the system — the compressors — were far removed from the guest rooms.

The indoor air handlers we used managed to be fairly attractive. They were 20 inches wide by 14 inches high and protruded out from the wall just 5 inches. (These can also be ceiling- or floor-mounted.) Each air handler has a small, quiet fan and is individually controlled, which means that the whole system is inherently zoned, which means that it saves the owners money, which makes everyone happy.

I can still hear Mary Van, with that fabulous southern drawl of hers, saying, "Ya'all did a fine job. Come back and stay at our B&B, ya hear?"

The most powerful split-system air conditioners on the market, manufactured by Dynazone, Enviromaster, and Klimaire, can run up to four separate air handlers. Some other makes, which can

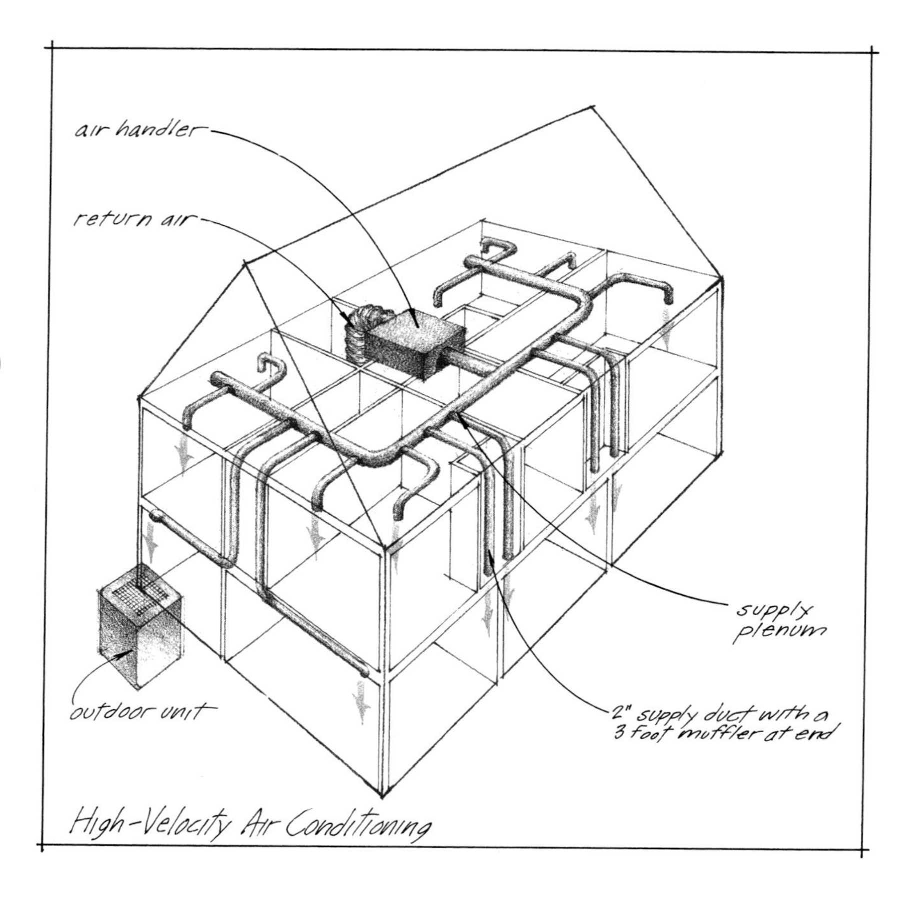

High-Velocity Air Conditioning

handle from one to three rooms, include Burnham, Freidrich, Toshiba, Hitachi, Typhoon, Sanyo, and Mitsubishi. Most of these manufacturers offer heat pump models, which means their systems can also provide heating.

The installed price of a split system runs from about $1,500 to $2,500 per ton. (A ton is 12,000 BTUs — the output of a midsized room air conditioner.) Like central air conditioners and heat pumps, the efficiency of split systems is reckoned in SEERs (Seasonal Energy Efficiency Rating), which range from 10 to 12.

Another economical way to retrofit an old house with air conditioning, which we've demonstrated twice on the show, is to thread flexible, pre-insulated ductwork down through the interior walls and closets. As with other types of air conditioning, the compressor and condensing coil remain outside, serving an indoor air handler.

As you can see from the drawing, the plastic flex duct, which is typically 2 inches in diameter, makes it possible to provide an air conditioning outlet to almost every room in the house, even when the house is two stories.

In both of the installations we did on "This Old House" — in Melrose and Wayland, Massachusetts, we positioned the air han-

Equipping the old house in Wayland, Massachusetts, with air conditioning required us to position the air handler up in the attic, supported from the rafters above (right). The air outlets on this type of high-velocity system (above) are relatively small and inconspicuous, so they don't divert attention from the home's beautiful wall frescoes and other fine features.

dler in the attic and mounted the room air outlets in the ceilings. In a different house, however, the air handler might have been placed in a closet or basement and the outlets mounted on the walls or even in the floors.

Because the plastic duct is small in diameter and corrugated (which impedes air flow), this type of system requires a much higher velocity fan than other air conditioning systems. Yet drafts aren't really a problem if the equipment is correctly sized and the outlets properly installed. As the airstream enters the room, it creates a gentle mixing of the air and eliminates stratification — that is, hot and cool layers of temperature. Warm air is returned to the air handler through a large return air grille and duct.

High-velocity air conditioning systems can also be designed to provide heating. This is accomplished in one of three ways: by using a heat pump; by equipping the air handler with a hot-water coil linked to a boiler or water heater; or by fitting the air handler with an electric heating element. Unico and Hydrotherm are among the companies that have pioneered high-velocity air conditioning.

Buying New

Regardless of whether your house is old or new, whether you're putting in air conditioning for the first time or replacing old equipment, the twin keys to a good installation are hiring a good contractor and buying properly sized equipment.

Just as oversized heating systems cause comfort problems and unnecessary expense, so too does air conditioning equipment that's too large for the job. An oversized air conditioner costs more both to buy and to operate, and isn't nearly as effective at dehumidifying indoor air as one that's properly sized. The only legitimate reason I can think of to oversize an air conditioning system is if you have *sure* plans to expand your home in the future.

So be sure that your HVAC contractor does a proper heat-gain calculation on your house. This involves the same steps that are required to size a heating system (see chapter 6), plus the added step of calculating the home's internal heat gains. Internal heat gains include all of the heat thrown off by appliances, electric lights, and people, which can add a lot to the air conditioner's load.

A heat-gain calculation will tell you precisely how many BTUs per hour your new air conditioning equipment needs to provide. No guesswork.

Here are some additional pointers to keep in mind when you go shopping.

- A good contractor should know as much about the control and distribution elements of the air conditioning system as he does about the central unit.
- To save you money, some contractors will recommend that you keep the indoor coil (provided it's in good shape), replacing only the outdoor unit. But this is usually penny wise and pound foolish. The indoor and outdoor coils need to be precisely matched for good performance. It's usually best to replace them both.
- Central air conditioning systems and heat pumps both carry an SEER (Seasonal Energy Efficiency Rating) to indicate their cooling efficiency. These range from 10 on the low end to about 17 on the high.

  Heat pumps carry a second rating — the Heating Season Performance Factor (HSPF) — which denotes the machine's heating efficiency. These range from 6.8 up to 10, depending on the make and capacity. The farther north you live, the more emphasis you should put on the HSPF side of the rating.

  Remember that it doesn't *always* make sense to choose the highest-efficiency equipment on the market. Why, for example, pay for an ultra-high-efficiency air conditioner if you're only going to use it a couple months out of the year and electricity only costs 6 cents a kilowatt hour? Conversely, if you're going to use the equipment a lot and/or have high electric rates, the argument for high-efficiency equipment becomes compelling.
- Some air conditioners that have high SEERs aren't very effective in removing moisture from the air. That's okay if

you live in a dry climate. But if you live in a humid area, talk with your contractor about selecting an air conditioner that has a high capacity to remove latent heat — that is, moisture. This may mean compromising the SEER rating a little so that you can get better humidity control.

Some air conditioners and heat pumps come with a two-speed compressor, which allows more flexibility in handling changing humidity conditions. Air conditioners with variable-speed compressors are even more flexible, combining very high SEERs with excellent humidity control.

A few state-of-the-art systems go a step further and use variable-speed motors on the blower as well. Using temperature and humidity sensors located inside the house, and a microprocessor control to run the show, these advanced systems can set precisely the compressor and blower speeds to deal with changing indoor-air temperature and humidity.

The energy payback on variable-speed motors, which can add $1,000 or more to the price of a system, can be quite long. But the extra comfort and quietness they provide are real advantages.

- As I explained in chapter 6, zoning a forced air system can provide powerful comfort and cost advantages. An intelligently zoned air conditioning system will deliver cool air where you need it when you need it, and avoid the waste of cooling empty rooms.
- Be sure to talk with your contractor about how your new air conditioning system can be designed to help meet your home's ventilation needs. Otherwise you could end up with a system that just recycles stale air over and over and over.
- Take care in deciding where to locate the outdoor unit, which contains the compressor. Noise is one important consideration. Another is performance. On a central air conditioner, the outside unit works best on the north side of the house, where it's shaded and cool. The same rule holds true for heat pumps that are used mainly for air conditioning. But in colder climates, a heat pump will heat more efficiently if it's positioned on the south side of the house, with some natural or man-made provision for summertime shading.
- Obviously, the placement of the indoor equipment is also important. It needs to be situated so that it's easily acces-

sible for servicing yet far enough away from living areas so that it doesn't create a noise nuisance. Also, your contractor should make sure that the air handler is equipped with a proper drain. An air conditioner can condense several gallons of water a day, as warm, moist indoor air comes in contact with the cold coils. If that condensate isn't properly drained away, you can end up with a little wetland in your backyard, replete with insects, alligators, and tropical birds.

- Most air conditioning systems come with a one-year warranty on parts and labor. The key component — the compressor — is usually covered for five to ten years.

  Thanks to the introduction of the scroll compressor, which has only three moving parts, I expect the warranties on compressors to get longer and stronger. You'll pay more for a scroll compressor than you would for a conventional piston and cylinder design, but they're quieter and longer lived, and in my view worth the extra bucks.
- Don't forget to check with your local utility to see if there are subsidies or rebates available for new cooling equipment. One utility I know of gives a $100 rebate on whole-house fans. Another pays $60 a ton if you select high-efficiency cooling equipment.

The Amazing Black Box

Without too much effort, I can imagine a day when all of the home's heating, cooling, ventilation, and water-heating needs will be handled by a single, compact, efficient, user-friendly machine.

Though it sounds like something out of "Star Trek," this amazing black box (or maybe it will come in designer colors) isn't as far out as some might think.

Already, the latest generation of air-to-air and geothermal heat pumps include models that not only heat and cool, but also provide hot water. It won't be long, I expect, before these so-called integrated or three-way heat pumps incorporate ventilation into their mix.

The key to building such a complex machine, and having it succeed in the marketplace, will lie in its ability to sense accurately what's going on inside our houses — temperature, humidity, air quality, human presence, and the operating status of other appliances — so that real-time information can be used to make discreet, automatic adjustments in our environment. Sensors will also become common on the outside of the house — measuring temperature, humidity, wind speed and direction, and air quality — so that

For More Information

Air Conditioning & Refrigeration Institute
Dept. U-181
P.O. Box 37700
Washington, DC 20013

- "Consumer Guide to Efficient Central Climate Control Systems"
- "How to Keep Your Cool and Save Cold Cash"
- "Heat, Cool, Save Energy with a Heat Pump"

American Council for an Energy-Efficient Economy
1001 Connecticut Avenue, NW
Suite 535
Washington, DC 20036

- *The Consumer Guide to Home Energy Savings* (Book: $6.95. Includes extensive listings of high-efficiency central and room air conditioners.)

Association of Home Appliance Manufacturers
20 North Wacker Drive
Chicago, IL 60606

- "Directory of Certified Room Air Conditioners"

the system's microprocessor brain knows where the balance points are between the outdoor and indoor environments.

So on a summer afternoon — a few years from now — this amazing black box of mine will be able to sense that my wife, Chris, and I are at home, and that the temperature and humidity inside are too high for us to be comfortable. It will check the outdoor air to see if it's suitable for natural cooling (no, too hot and humid) and quietly turn on the air conditioner. Only a select part of the first floor will actually be air conditioned — "Zone three," according to the machine — where Chris and I (in a nostalgic mood) are watching a twenty-five-year-old videotape of "This Old House." Without our ever thinking about it, the "waste" heat coming off the air conditioner's coil will be routed automatically into the water heater. A few hours later, my smart and friendly machine will realize that the outdoor air has cooled down to the point where it can now be inducted beneficially into our house and it will quietly turn off the air conditioner and let nature — with a little fan to assist her — do the nighttime cooling and ventilating.

A dream, you say?

Maybe. But I wonder what my great-grandfather Harry — an old-time plumbing and heating man — would say if he could lay eyes on some of these modern-day heating and cooling machines.

In the end, who knows what technology and time will conspire to bring us. Could the Wright Brothers, peering forward from their perch at Kitty Hawk, have ever imagined a 747?

A Word of Thanks

A lot of good friends and colleagues have helped us in preparing this book. So too have a lot of private companies and public organizations. To all of these individuals and entities, who have been so generous with their time and information, we tip our grateful hats.

Academy of Motion Picture Arts; Air Conditioning and Refrigeration Institute; American Gas Association; American Solar Energy Society; A. O. Smith, Inc.; Austin Power & Light; Jack Banker, Rheem Manufacturing Co.; Perry Bigelow, Bigelow Homes; Steve Bliss, *Journal of Light Construction*; Bob Block, Sage Advance Corp.; Boston Edison; Broan Manufacturing Co.; David Brown, Princeton Energy Partners; Carrier Corp.; Subrato Chandra, Florida Solar Energy Center; Rick Cowlishaw, AIA; Paul Cronin; The Earth Works Group; Edison Electric Institute; Helen English, Passive Solar Industries Council; Richard Faesy, Energy Rated Homes of Vermont; Gas Research Institute; Doug George, Doug George Homes; Dr. Michael Holick; Honeywell, Inc.; Bion Howard, Alliance to Save Energy; Ron Hughes, Energy Rated Homes of America; Hydronics Institute; Joe Iorio, Atlantic Heating and Air Conditioning; Dennis Jaasma, Virginia Polytechnic Institute and State University; Paul Knight, author of *Mechanical Systems Retrofit Manual;* Dr. Rich Krajewski, Brookhaven National Laboratory; Lennox, Inc.; Madison Gas & Electric Co.; Ed Mantiply, U.S. Environmental Protection Agency; Massachusetts Audubon Society; Merle McBride, Owens Corning Technical Center; Phillip C. McMullan, Thermoscan Inspections; Alan Meier, *Home Energy;* Tom Mooney, Retrotec; Mor-Flo Industries; National Fire Protection Association; Northeast Utilities; Jason Perry, Vermont Castings; Tony Poanessa, Sunrooms, Saunas, and Spas; Rocky Mountain Institute; Philip Russell, Russell Home Builders; Hank Rutkowski, HTR Engineering Services; Larry Sanford, Macklanburg-Duncan, Inc.; Southwall Technologies; John Spears, Geomet Technologies; Dr. Richard Standwick, City of Winnipeg Health Department; State Industries; Steven Strong, Solar Design Associates; Earl Vaught, Taylor Environmental, Inc.; Frank Vigil, North Carolina Alternative Energy Corp.; Dr. Lance Wallace, U.S. Environmental Protection Agency; Al Wasco, Jim LaRue, and Don Jones, Cleveland Housing Resource Center; Larry Weingarten, Elemental Enterprises; Bede Wellford, AirXchange; and Bob Wood, Hurd Millwork Co.

Richard Trethewey & Don Best

Photography Credits

2
Photography: Richard Trethewey

3 (top)
Photography: D. Randolph Foulds
Country Home magazine, October 1990. © Copyright, Meredith Corporation. All rights reserved.

3 (bottom)
Photography: William Stites
Country Home magazine, October 1990. © Copyright, Meredith Corporation. All rights reserved.

4
Photography: Richard Howard

13 (top)
Photography: Neil Jacobs
courtesy "This Old House"/WGBH

13 (bottom)
Photography: Marty Snortum
courtesy "This Old House"/WGBH

23
Photography: Richard Howard

29
Photography: Richard Howard

33
Photography: Bill Schwob/WGBH

39
Photography courtesy Energy Rated Homes of Vermont

57
Photography courtesy TSI, Inc.

59
Photography courtesy Boston Edison

66
Photography: Scott Baxter/WGBH

67
Photography: Michael Lutch/
WGBH

68
Photography: Michael Lutch/
WGBH

69
Photography courtesy Four Seasons Sunrooms

70
Photography: Norman McGrath
Design: Peter Gisolfi Associates, Hastings-on-Hudson, NY

71
Photography: Norman McGrath
Design: Peter Gisolfi Associates, Hastings-on-Hudson, NY

74
Photography: Michael Lutch
courtesy "This Old House"/WGBH

77
Photography courtesy Doug George Homes

78
Photography courtesy Rick Cowlishaw

79
Photography courtesy Russell Homes Builders

89
Photography: Bill Schwob
courtesy "This Old House"/WGBH

126
Photography: Richard Howard

130
Photography: Richard Howard

243
Photography: Richard Howard

Index

Y

Z